A COLD WAR SAILOR

BY

J. J. COYNE

CAPT USN (RET.)

Book Cover Design and Publishing Assistance Provided by Amazonlistinghub

DEDICATION

To Carey, Kelly, Erin, and especially their Mom, Linda

None of this would have been possible without you!

<u>Acknowledgement</u>

This memoir portrays events to the best of my recollection. Any errors are mine alone. I can never thank all those with whom I served that made me a better aviator, officer, and man. A special thanks to my family who encouraged me to pursue my dreams.

TABLE OF CONTENTS

CHAPTER I:

SO, YOU WANT TO BE A NAVY PILOT?

I looked up from the instrument panel in my Huey helicopter to see the main rotor blade strike a tree.

The next sensation was one of falling, hitting the side of the hill in the Talladega National Forest in rural Alabama, followed by a sudden stop. Well, dying was not as bad as I thought it would be! Now, if someone would only turn off that jet engine screeching in the background. Wait a minute, I am not dead, but if that engine is not turned off, it could explode, and the whole helicopter would go up in flames. Great, survived the crash but died in the fire.

Welcome to the world of Naval Aviation.

I tried to shut the engine down with the twist grip, but it would not respond. A sound and movement caught my eye. I realized the instructor pilot was asking me if I was ok. He had exited the aircraft from the left and come around the nose to see if I was alive. I nodded yes, and he moved aft, and moments later, the sound of the jet engine stopped.

The aircraft had come to rest on the side of a hill. I released my shoulder harness and fell forward into what was left of the pilot's compartment. As I picked my way out of the wreckage and joined the instructor pilot and a fellow student who had been in the back, we tried to figure out what had happened. None of us was injured. Although looking at the wreckage, it did not seem possible that at least one of us had not suffered a major injury.

Within a few minutes, we heard a man call from below, "how y'all doin'? Heard you comin', and then I didn't hear you no more, so I figured I would see how y'all were." We made our way back to the farmer's house and called all the appropriate authorities to advise them of our situation, assured them we were all fine, and told them not to call our wives.

Later at Fort McClellan, we went through a medical examination, had blood drawn, received a medical thumbs up, and turned our attention to quarters for the evening and adjourning to the Officer's Club for a steak and a drink.

Taking the opportunity to call home, my wife Linda answered with a question, "What happened?"

I told her we had been in an accident but that I was ok and it would be a few days before I would come home. The daughter of a Naval Aviator, she said, "How bad is it?" I was channeling Chuck Yeager as I replied, "It's a little bent up." The duty office had called earlier, and she said the conversation went something like this, "Ma'am, your husband has been involved in an aircraft accident." There was a pause on the line, which she later described as the longest two seconds of her then-young life -then the petty officer came back on the line and said, "But he is alright and will call you when he gets a chance."

Navy aircraft accidents, especially when an aircraft has been destroyed, require convening two investigations that occur simultaneously.

The first investigation is the Accident Mishap Board or AMB. Its' purpose is to determine exactly what happened and how to keep it from happening again. The AMB members are fellow pilots and focus on whether the accident occurred due to pilot error, mechanical malfunction, procedural error, weather, etc. By the time my Navy career was over, I would be the subject of an AMB, serve on an AMB, and convene an AMB.

The second investigation is the Manual of the Judge Advocate General or JAGMAN. A JAGMAN is a legal investigation to decide whether the accident occurred in the line of duty or if some rules or regulations were violated.

The AMB asks you, "Can we talk?" and the JAGMAN reads you your legal rights.

An AMB was convened in Alabama. The arrival of the members of the AMB was like a scene out of Apocalypse Now. It seemed like there were Huey helicopters everywhere, and I very rapidly found myself in a room with a group of instructor pilots who began to ask me questions about the flight. As a Student Naval Aviator (SNA), Instructors are Gods, and like Gods, they hold your fate in their hands. I was at the very end of the flight syllabus and could see those wings of gold beckoning in the distance, and I did not want to screw that up by not giving the correct answers.

The flight was scheduled as a low-level cross-country on a Saturday, and the weather was "dog crap," a highly technical term that denotes slightly above to slightly below the minimums required to conduct the flight. The weather guessers said it would get better during the day, the rain showers would end, and the clouds or ceiling would rise while we were flying the hop.

They were wrong, the ceiling lowered, and our flight took us across rising terrain. Can you see what is coming?

Although I was supposed to navigate the low-level flight, I was lost. As I shuffled the map, I looked over, and the Instructor Pilot (God) seemed to know where we were. When I confessed that I did not know where we were, he flew on for a few minutes and then announced we were going to land. He got out of the aircraft and went up to a farmer's house. About five minutes later, he returned and told me where we were and that we would hang a right, follow county road so and so to Talladega, and then turn right when we intersected I-20, which would take us to Anniston, Alabama, and Fort McClellan.

We had made the turn and were following the county road which was in a valley. The rising terrain and the lowering ceiling now raised their ugly heads. The Instructor Pilot was flying, and I was looking outside to keep clearance on my side of the aircraft with the valley hillside. As I was doing so, we went into the Goo, another highly technical term for inadvertent instrument flight, like being in the middle of a cloud. The road was beneath us when I looked down through the chin bubble. Then I looked outside and saw nothing but the cloud. About this time, "God" in the other seat announced, "I have vertigo. Back me up on the controls!" We were both on the controls, and I looked at the attitude gyro, which was nose high and right wing down. I initiated a correction and looked outside for a visual reference. Nothing but clouds! Back inside, the attitude gyro was still not level, and I tried to correct it again. Of course, God and I are counterbalancing one another as we wrestle with this beast in the Goo. About then, I looked outside and saw the main rotor blade hit the tree. You know the rest of the story.

One of the AMB members recounted all the things that happened during the flight, marginal weather, getting lost, landing in a farmer's field, going inadvertent IFR and posed the following question to me, "at what point did you realize that things weren't going well on this flight?" My response was immediate and struck the room like a thunderclap, "when I saw the main rotor blade hit the tree, Sir!" The Navy knew then they were not getting the sharpest knife in the drawer.

I had not realized how serious the accident was initially because we all had come through ok. It began to dawn on me that this was not the norm when one of the AMB members asked me which seat I had been in during the flight and just shook his head in disbelief. When the aircraft wreckage arrived via flatbed truck back in the squadron, it began to sink in that I was an incredibly lucky man.

No one ever briefed me on the results of the investigation. I started flying when we returned to Whiting Field in Florida, and I received credit for the low-level flight during which the accident occurred. My next hop would be confined area landings which required landing the aircraft in a small field in the middle of some very, very tall trees. It took all the weight of the Instructor Pilot on the collective to get me to descend into that field. Two weeks later, I had my wings.

UH-1 CRASH IN TALLEDEGA NATIONAL FOREST

Three years later, while deployed on detachment, I got into a discussion with LT Dave Youngblood, my Officer-in-Charge (OINC), about our most harrowing flight experiences. He had a couple of tours in the fleet and had been an instructor at the Flight Instructor Training Course (FITC) in Pensacola. That course was designed to prepare instructor pilots for all the crazy things their students would do to them during their flight training. My accident was one of the case studies Dave briefed. He said he would ask the instructors what happened to the aircraft commander after this accident. Some said he must have lost his wings; others that he had been court-martialed. Dave paused for effect and then told the gathered instructor pilots that the aircraft commander was now flying Fat Albert with the Blue Angels.

Even after all these years, when asked the same question, "When did you know things weren't going well?" My response is always the same. “When the God@##$ main rotor blade hit the friggin’ tree!!!”

CHAPTER II

PETER JENNINGS TO NAVAL AVIATOR

I did not start out to become a naval aviator. In fact, I wanted to be the starting quarterback for the Cleveland Browns. By high school, I knew that was not going to work out, so I shifted to the idea of becoming a broadcaster. I could anchor the network news like Peter Jennings or do play-by-play for the Browns instead.

Going to college was going to be a stretch financially. The second of five kids, the son of a police officer and homemaker, I looked around for an in-state school that had a good broadcast journalism program and narrowed the choices down to Ohio State University (OSU) and Ohio University (OU). I still had some athletic ambitions and figured I could walk on easier at OU than at OSU, but I soon figured out I better stick to communications (Radio-Television).

OU provided me with unlimited opportunities to gain experience in the broadcasting journalism business. My financial package consisted of what my parents could afford, a small grant, National Student Defense Loan, and a work-study program. My work study was at the university radio station WOUB where I initially focused on the technical side of the business instead of the on-air or "talent" stuff. I learned how to operate the board in the studio, how to set up and run all the equipment, and progressed to setting up the telephone line and all technical equipment necessary to do a remote broadcast for high school football and basketball games. I got so good at it that they sent me along to help with some of the football road games for Ohio University. That led to doing the setup and working home OU basketball games. All this experience prepared me to step out from behind the scenes and become a "talent."

As a part of my work study, I was asked if I would do the classical music show on Sunday nights on WOUB. I shudder to think about how I "murdered" the pronunciation of the composers and their works. I was more at home doing the "Town and Country" report at noon every day. You learn a lot about the price of pork bellies and future grain prices on a show like that. When you are starting out, you make sure you read all the copy before you go on air. That way, you are not surprised by a name and can get the phrasing and text just right. However, over time, and when things change in the studio, especially breaking news, you must read the copy "cold" or not have previewed it. One day on the Town and Country report, I was going to read some copy about pest control between records and had not really looked at the copy before I was going to read it. I decided to ad-lib a little bit before the record and said, "The best way to hunt ants and make sure you kill all of them is to spread some powder on the floor and then track them down with a sledgehammer. That way, you can't miss. I paused a moment and then started to read the copy. It said to spread powder on the floor so that you can track the ants. I lost it and started laughing, could not stop, and just started the next record. The program director was listening and was not happy with either the ad-lib or losing my composure on the air. I never did a cold copy read again.

Coming from Cleveland, I did not realize how important these smaller-town high school teams were to their communities. Everyone listened to the broadcast of their games on the weekend. We traveled all over Southeastern Ohio doing Athens High School games as a member of the Southeastern Ohio Athletic League. Going into my sophomore year, I began to do play-by-play for high school football and basketball. I had a face for radio, so I learned all the technical positions for television and started working camera, audio, lighting, assistant director, etc., for the nightly news program. As with radio, by my sophomore year, I transitioned out from behind the camera to do news and sports a couple of nights a week as a talent in front of the camera on WOUB Television.

There was a line that separated the news folks from the sports folks, and I was the only one who did both during my time at OU, both on radio and television. The news folks did not view the sports folks as real journalists. I showed the News Director that I could do both. By my senior year, I was providing color commentary for the Ohio University football network and doing play-by-play for the basketball team. During my sophomore and junior years, I did play-by-play for high school football and basketball. Like most folks in college, I had found my niche and spent all my time involved in broadcasting and hanging out with my fellow radio-television majors. By my senior year, I was doing radio play-by-play on the weekends for high school games and color for the OU football network. I was also doing three nights a week on television, two for news and one for sports.

All that experience allowed me to get a summer internship going into my senior year at WVIZ, the Public Broadcasting Service television station in Cleveland. As a booth announcer in the evening, I wrote and produced promotional program spots, did station breaks, and provided emergency weather notifications. Before I left the station to go back to school for my senior year, the Program Manager offered me a job and said he was willing to keep a slot open if I would commit to returning after graduation. I turned him down. If I returned to WVIZ, I would be doing the same things I had done as a student over the summer as an announcer behind the camera. I wanted to do news and sports in front of the camera. He could not guarantee me a spot as "talent" if I came back to WVIZ, but he felt sure a position would open up over time. Months later, I would regret the decision to turn the job down.

When I graduated from college in December of 1971, the job market was not great, especially in broadcasting. After moving home, I eventually got a job as a service representative for a Volkswagen dealership in a Cleveland suburb. While I continued to send out resumes and network with my peers, there were no jobs available in broadcasting. Even the job at WVIZ had been filled. This was at the beginning of 1972, and Vietnam was still going on.

When I went to college in 1968, the draft was in full swing, and I figured if I must go into the military, I would rather go in as an officer. I joined the Army Reserve Officer Training Corps (ROTC) program at OU in 1968. When they held the first draft lottery in 1969, my number was 357. I figured it would have to be a nuclear war before I was drafted. I promptly resigned from ROTC. So, after graduating and after about six months of working for the Volkswagen dealership, I started to reconsider my options for going into the military. I decided to see the Navy Recruiter about what he might have to offer.

My dad had been a Chief Petty Officer in the Navy. He enlisted in 1939 and became an Aviation Ordnance Man on the USS RANGER, CV-4, the first U.S. ship designed and built to be an aircraft carrier. He spent the first part of World War II on RANGER in the Atlantic but then received orders to Rhode Island for shore duty. The needs of the Navy soon intervened. His shore duty was terminated, and he was reassigned to the USS SAN JACINTO, CVL-30. Most folks recognize that name as being associated with former President George H.W. Bush, but to me, it will always be my dad's ship. Several of my uncles were battleship sailors on USS NEW MEXICO and USS MISSISSIPPI. I grew up listening to the exploits of the fast carrier task forces that paved the way for victory in the Pacific. I heard that special tone in my dad's voice when he talked about the Naval Aviators. In the back of my mind, I wondered if I had what it took to land aboard a ship.

I took the Armed Services Vocational Aptitude Battery (AVSAB) test and figured I would get a call back from the Navy immediately. Turned out that I did not do that well on the test. Eventually, the Navy called back and said, don't call us. We will call you. A few months later, I got a call from the Navy saying that I had qualified for the Naval Flight Officer (NFO) Program, and they offered me a slot. Simultaneously, a broadcasting friend called to say that there was a job at the local radio station in Logan, Ohio, if I was interested. Deciding to take the job in Logan, I called the Recruiter to thank him for the offer and explain why I had to give broadcasting a shot. He said he understood and encouraged me to write a declination letter that would leave the possibility of going into the Navy later if I changed my mind in the future. I did so and left for Southeastern Ohio.

I was hired at WLGN to be the News Director, do the afternoon radio show, and play-by-play for the Nelsonville York Buckeye's high school football team. While I loved the folks I was working with and enjoyed the community, the money was not that great. At the time, I was thinking about getting married, and the commercial radio station in Athens, Ohio, WATH, offered me a larger salary to be their News Director and do play-by-play. Again, I enjoyed what I was doing and the folks I was working with, but it became clear to me that job security in the broadcasting business was not that great and that, in general, you got a raise by moving to another city and station.

The alternative offer by the Navy began to look more attractive, although I did not want to be a Naval Flight Officer. I purchased the AVSVAB exam practice book and began studying for the test. The second time was the charm as I qualified for selection to enter pilot training through the Aviation Officer Candidate School (AOCS) and reported to Pensacola on September 11, 1973, to join Class 3773.

CHAPTER III

ROCKY START

AVIATION OFFICER CANDIDATE SCHOOL (AOCS)

SEPTEMBER – DECEMBER 1973

Upon arrival in Pensacola, I was immediately impressed by the heat and humidity. Some friends from college picked me up at the airport, and we went out to dinner. Jim had done a tour on a destroyer and was now training to be an intelligence officer. He talked about life in the surface navy and regaled me with stories about liberty in the Mediterranean. He said the ship's commanding officer had an enormous impact on the ship's morale and combat efficiency, something I would experience up close and personal later in my career.

Jim dropped me off at the Quarterdeck, and I went inside to check in. The officer of the deck said he could not check me into the Schools Command that late in the evening. He told me to go to the transient barracks and return for check-in the next morning. The seaman behind the desk at the transient barracks said he did not have any sheets or covers and laughed when I asked about a wake-up call for the next morning. That night was spent worrying about oversleeping, upset at the way I had been treated, and beginning to wonder if I had not made a big mistake.

A drill instructor (DI) stood in front of the Aviation Officer Candidate School (AOCS) Indoctrination Battalion and urged a couple of us who had gathered there to "come on over!" Behind him, several other DIs were strenuously encouraging (yelling) at those who had accepted the offer and inquiring whether they had any weapons or drugs in their possession. The next few days were a blur of paperwork, haircuts, issuing uniforms and other gear, measurements, medical examinations, and a lot of yelling. Army ROTC at Ohio University had prepared me to shine my shoes and brass, but not for the focused attention of a Marine Drill Instructor.

We adjusted quickly to the rhythms and routines of AOCS, although for about the first three weeks, I was beginning to wonder if I had joined the Marine Corps by mistake. The DI got us up in the morning, marched us everywhere we went, including chow (meals), ensured we had enough exercise, and then put us to bed. I caught on rapidly that there were no correct answers to any question you were asked, and whatever you said led to a lot of exercises. So, the safest course of action was to say nothing and then take your punishment. "Yes Sir, No Sir, No excuse, but I will find out Sir," was seared into my brain. I learned about preparatory commands and commands of execution. Such as a preparatory command of "Do side straddle hops (jumping jacks) until you have one foot in the grave" and the command of execution, "Ready, begin!" I marveled at how slippery the passageway (hallway) could become with sweat when forty-plus young men were doing pushups, jumping jacks, and running in place in the delightful Pensacola fall weather.

A day in the life of an Aviation Officer Candidate began with reveille at 0500, and the rest of the day combined academic instruction, military drill, physical training, meals, study, and sleep. You learned the meaning of teamwork and depending on your classmates. If you did something particularly well, you would teach your classmates how to do it. You held the rank of E-5 if you were a college graduate going through the program. If you were prior enlisted, and at least half of my class had been, you were paid at your previous rank, with longevity for prior service. The guys with prior military experience helped those who did not, and those with a technical degree helped those with liberal arts degrees learn about aeronautics, mechanical engineering, and meteorology.

The Drill Instructor got you up in the morning, marched you to and from class, took you out for nice long cross-country runs, and "motivated" you on the obstacle course. The DI would come through the class area at night when you were studying to make sure that you were in your own room, focused on your study, and wish you a pleasant evening as taps were sounded and the lights were turned out. He would ask if you had any "pogey bait," sweets, cookies, or candy. Our Drill Instructor's Military Occupational Specialty was as a baker. He would sample any baked goods that were sent to you and offer his expert opinion as to how good they were. Our nickname for SSGT W. D. Perry, USMC, was "Cookie Monster."

The most significant or lasting lesson I learned was in leadership class. The instructor gave us three principles to guide us as naval officers. He told us that we needed to "Be a man, know your stuff, and take care of your people." It was pretty simple, but it struck a chord, and I would use them as the basis for my leadership style for the rest of my career.

"Be a man." It was obviously a different Navy. The first female aviators were beginning flight training when I reported to Pensacola. I did not take it literally. Rather, to me, it meant having honesty and integrity. Stand up for your principles, speak truth to power, and accept responsibility for your actions and those who worked for you. Do the right thing, and if that is not clear to you, then do no wrong. Acknowledge your mistakes, learn from them, and move on.

"Know your stuff." I translated that into being a professional. Make yourself an expert on whatever your duties and responsibilities were. As an aviator, know your aircraft inside and out. How the systems worked, limitations, normal procedures, emergency procedures, and tactics. Never waste time on a hop. Make sure you focus on and practice the hard things or things you do not enjoy doing. Imagine the worst thing that could happen to you in the airplane and then train so that you could manage it with ease. As a naval officer, the important thing about the Navy is that everyone has a billet or job to which they are assigned. Every billet has a description of the duties that you are responsible for, the governing instructions or publications that apply to that billet, and for whom you work. All of these are resources that prepare you to perform in that billet to the best of your ability. Know your stuff cold and constantly strive to get better.

"Take care of your people." As a former Chief Petty Officer, my dad had drilled into me the tenant that if you take care of your people, they will take care of you. I learned that sailors did not want to be coddled or babied. They demand respect and that you treat them as adults. A division officer's job is to ensure that your sailors have everything they need to prepare them for success. That their basic needs: food, shelter, and pay, are being met. That they have the proper school, training, tools, parts, and equipment needed to do their jobs. If they had an administrative problem with their pay, I would go to disbursing to straighten it out. I figured I would get better and faster service than them, and I wanted them to be able to focus on their job.

While I always said that the mission took priority over the people, I knew that without the people, the mission would never be accomplished. You learned over time to provide the objective to be accomplished and then let your sailors figure out what was required and the best way to accomplish it. Liking me personally was not important. I did not care if they liked me. I learned about different personalities and what types of motivation to use depending on the individual and the situation, i.e., the pat on the back instead of the kick in the ass. The best thing in the Navy is the people, and the worst thing in the Navy is the people. If you are not careful, you will spend most of your time on the bad sailors and take the good sailors for granted.

There were also a surprising number of funny moments.

We were about halfway through the program when we returned from class one day to find our rooms had been ransacked, racks overturned, and cleaning powder all over the floors. We all crowded into the room of our class leader. Four officer candidates would share a room. If any of them dropped on request (DOR) from the program, their locker was emptied, and their rack was stripped. There was an empty rack and locker in this room, which allowed even more of the class to crowd in. Everyone was complaining about what had happened, how the DI had it in for us, and bitching about the Marine Corps in general. There was a sound in the room, and we all stopped talking for a moment. A minute later, we started complaining even louder when we heard the sound again. It was coming from the "empty" locker. One of us went over and opened the door. Out stepped Staff Sergeant W. D. Perry, United States Marine Corps, the Cookie Monster. "37", our class number, "out in the passageway!" Forty guys tried to get out of the door simultaneously. We assumed the position of attention against the bulkhead (wall), but not before turning our highly polished belt buckles around so they would not encounter the cleaning powder on the deck, which we were about to become intimately familiar with. Thirty minutes later, you could not stand in the passageway because the floors were wet with cleaning powder and sweat. Staff Sergeant Perry had made his point we were not as good as we thought we were, and he would let us know when we were.

One day in the mess hall, we were just beginning to enjoy our lunch when they brought in the latest class of "poopies." That was a reference to the uniform you wore, a poopy suit, and the fact that you were the low men on the totem pole in your first week of training. The Drill Instructors were yelling and screaming at them to move along, maintain the proper distancing, and asking them questions about their chain of command, the general orders of a sentry, and a description of the M-1 Garand rifle. They were paying special attention to the first guy in the line. Nothing he did was correct, and they just kept harassing him. One of the guys at our table did a double-take and said to us, "Do you recognize the guy the DIs are yelling at?" We took a second look and were shocked. It was Staff Sergeant Sable, USMC, dressed up as a "poopie." He was going to be this new classes Drill Instructor. We all knew that when they got back to their barracks, they would all bitch about the Marine DIs, with one of the DIs sitting right there with them. We could only imagine the looks on their faces when they were getting ready to outpost from the Indoctrination Battalion to their regular battalion. Their drill instructor would be introduced, and out would come Staff Sergeant Sable. If I had been in the room with Staff Sergeant Sable for that first week of Indoctrination, I would have DOR'd on the spot.

About three weeks into the program, I was selected to represent the class as a member of the flag pageant. The flag pageant was a part of the commissioning ceremony that took place every week when a new class was commissioned. The good news was that it took place immediately following the commissioning ceremony on the parade grounds, where all the officer candidates had passed in review. Each week the three battalions were judged on how well they executed the manual of arms and marched in precision on the parade ground. The winner received praise, and the other two battalions were "rewarded" with extra drills and physical training.

Unless you were in the flag pageant, in which case you missed all the "fun."

The selection process for this was quite rigorous. "37 out in the passageway!" bellowed Staff Sergeant Perry. When we were all assembled, he said, "when I call your category, get back in your room. Sub-academics, get in your room. Sub-physical training, get in your room. Sub-military, get in your room. You three are in the flag pageant." I proudly served as the Rattlesnake Flag and constantly reminded my classmates not to tread on me.

The other good deal was I got to perform as a part of the flag pageant in Kingsville, Texas. Prior to leaving, we had a room, locker, and personnel (RLP) inspection to decide whether we would become "secured" candidates. Being secured was important. A secured candidate could talk during meals, have dessert, go to the exchange by yourself, and most importantly, go on liberty out in Pensacola, or in my case, see my wife. Everyone knew that you were secured by the pieces of tape that were on the ends of your nametag. No tape: you were a new guy. Tapes: and the world was yours.

The RLP was very intense. The class officer and the DIs would storm into the room. Get in your face and order you to recite the chain of command, general orders of a sentry, inspect your locker, describe the M-1 rifle, etc. The fold of the sheet on the top of your rack (bed) had to be exactly six inches, the same length as your folded tee shirt. The inspector would take your tee shirt out of the locker and lay it on the folded sheet. If they did not match exactly, he would ask you, "which one is correct?" If you said tee shirt, your rack ended up on the deck (floor). If you said rack, the drawer with your tee shirts ended up on the deck. By the time you had your RLP to become secured, you used a master tee shirt for the room so that everything matched up for you and the other guys in the room. The worst thing that you could be accused of was screwing your buddy. If one guy was not squared away, it was incumbent on the others to help him out. If your stuff was perfect and theirs was not, then you were screwing your buddy. By extension, if anyone in the class was having difficulty, it was incumbent on the class to help them out. Peer pressure was a constant factor. I departed for Kingsville, not knowing if the class had passed the RLP or not.

At Kingsville, I stayed overnight with a family that had volunteered to host me for the evening. They were genuinely nice, but it was strange to be away from the regimented confines of AOCS. It felt completely different, and I was struck by just how "military" I had become in just a few months.

Upon returning from Kingsville, I saw that my classmates were wearing their "tapes," and I was overjoyed. I started to make a beeline to our room so I could change my uniform and call my wife about coming to pick me up. When I got to the room, my locker was empty. All my clothes were gone, which was bad, but the worst part was that my M-1 rifle was missing. I had forgotten to lock my locker when I left for Kingsville. The DI had discovered it, trashed all my stuff, and took my M-1. The good news was that I could borrow a uniform from my roommates and go on liberty. The unwelcome news was that I would have to see the DI about getting my rifle back on Monday. I had heard stories about this happening. The DI would break the weapon down and give the pieces to other drill instructors. Then you would have to see each of them and do physical training to get the pieces back.

The office entry, also known as going to see the DI, was something that every candidate dreaded. You would center yourself in the hatch (doorway), pound the pine, sound off, and then enter the office. Sounding off consisted of the following, "Sir, candidate officer Coyne, JJ, Class 3737, request permission to see the drill instructor! This was recited as if it were one long word. If you were lucky, your drill instructor would answer, and you would enter, square your corners, and come to a position of attention exactly six inches from his desk, with your eyes locked on the bulkhead (wall), not at him. One of the DIs had a poster of a particularly buxom young woman behind his desk with the caption, "are you eyeballing me, candidate?" Eyeballing, moving your eyes at all when they were supposed to be locked straight ahead, was a capital offense. Gunnery Sergeant "Mad Dog" Delaney ordered me to enter, and I squared off in front of his desk. I informed him that I wanted to see Staff Sergeant Perry, and he asked me if that was the case. What was I doing in front of his desk? He told me never to darken his doorway again and to get out in the passageway and do physical training (PT) until there was nothing left of me but buttons and a pool of sweat. After about five minutes of that, he shouted out and inquired if I was prepared to do a proper office entry. I repeated the process; this time, Staff Sergeant Perry told me to enter. I did not have to PT for every piece of that rifle, but by the end of that day, I was exhausted and vowed never to leave the room again without checking my lock and, of course, all my roommate's locks.

After about eleven weeks, we became Candidate Officers and oversaw running the Regiment, which consisted of four battalions. There was a Regimental Commander who wore five bars on his uniform and was responsible for the entire Regiment, and four Candidate Officer Lieutenant Commanders who wore four bars and were responsible for running the First, Second, and Third Battalions and the Indoctrination Battalion. I was put in charge of the Indoctrination Battalion, also known as the Poopy Battalion.

Going back to the Indoctrination Battalion was strange as I would now be working with the DIs. However, I was also committed to trying to teach the Poopies what they needed to know to make their transition as easy as possible. I remember calling them together and laying out what was going to happen to them over the next week. I outlined all the ways the DIs would try to trick them as they went through the various evolutions and stressed that they needed to act together as a class. Ironically, they would get into more trouble doing things wrong as an individual than they would as a class. They did not believe me, and so I watched them get hammered for doing the things that I had expressly warned them not to do.

I was honored to lead the "Poopy" Battalion for two weeks over the Thanksgiving Holiday, but I could not wait to report to Saufley Field for flight training. When the two weeks were up, I left my home in Poopeyville, packed my trash, and reported to Saufley Field for initial flight training. We had four weeks left until commissioning, so we were still under observation by the Drill Instructors, but the daily firsthand stuff was greatly diminished. Our plates were full with just trying to adjust to becoming flight students, so they left us alone.

AOCS CLASS 37-73 NAS PENSACOLA

THE AUTHOR IS IN THE FIRST ROW, THIRD FROM LEFT

CLASS OFFICER LTJG M.G. McFee (Left) and SSG W. D. Perry (Right) in FRONT ROW

Checking into VT-1 was like being on the wrong end of the fire hose, but it was something you would learn to expect with every squadron you ever checked into. You were issued flight gear, academic textbooks, standard operational procedures, and flight schedules and instructed to learn an impossible amount of information by the next day. Lessons learned from AOCS, cooperate to graduate, were once again clear as classmates would work with their peers on areas that they needed help with.

Index cards with vital information on normal procedures, emergency procedures, airframes, engines, and limitations were produced and distributed. As we began to fly, the gouge (information) on individual instructors, or what their preferences and habit patterns were, would be passed among class members. Who was a "screamer" or who was a "good guy" would become common knowledge. Seeing your name next to a screamer on the flight schedule made for a sleepless night. In our room at Saufley Field, on the floor was the layout of the airfield and its runways. You could fly the landing pattern in the room using the floor layout before you had to do so in the air. We would go together to the squadron simulators to learn where all the gages, circuit breakers, levers, and throttles were. You would close your eyes, and your classmate would identify a gage or circuit breaker, and you had to reach out and touch it without opening your eyes. You would also be quizzed by your classmate on emergency procedures and aircraft limitations while doing the blindfold check. We even practiced bailing out the side of the aircraft into a mattress, so we could do it correctly the first time.

My first flight was more of a terrifying experience than an inspiring one. I was trying to remember all the things I was supposed to do and the correct sequence in which to do them. Also, I was busy trying to figure out where I was in relation to the earth, something that I would continue to have issues with (see helicopter crash in chapter one). Fortunately, the love and fun of flying would come later despite my being an average pilot.

The two things that began to consume me were soloing and being commissioned. We had reported to Saufley Field the first week in December, and it soon became apparent that the Christmas schedule would affect how quickly we made it through the initial part of the syllabus. In the past, you had to complete a solo flight prior to being commissioned. That requirement had been done away with, but if you received a "down" or an unsatisfactory flight performance, then your commissioning date could be pushed back.

I wanted to be commissioned with my class more than anything. I stayed in Pensacola over Christmas and was able to make my way through the syllabus quickly. When my classmates returned from Christmas leave, they had to complete a couple of warm-up hops. The weather in Pensacola made it hard to get the two warm-ups before they could continue in the syllabus. But I had been flying all along, so no warm-up was necessary, and all my hops counted against the syllabus and completion of the Pilot Training Requirement or PTR. I would be launched with an instructor to do the weather reconnaissance to figure out if the squadron should launch the students. Often, that might be the only hop to get out that day. So, I became a priority for flight scheduling which meant my chances of getting a down and not being commissioned with my class were increasing. Add the fact that my parents were coming down for commissioning, and the stress factor was off the charts.

Commissioning was scheduled for Friday, and I took my safe for a solo check ride on Wednesday. Most of my classmates were not flying that week. Some of their instructors had encouraged them to develop "sinus" problems during that week so that they would not run the risk of getting a down. I was just too stupid to think about doing that. When I went out on the check ride, I thought I had failed it. It seemed that everything I did was wrong. I fumbled my way around the pattern to some barely survivable landings. When we got back, the instructor left me in the debriefing room and went down the passageway. I figured he had gone to tell the flight leader that the PTR was about to take a hit, and Coyne had gotten a down (unsatisfactory) on his safe for the solo check ride. After what seemed like an eternity, he came back and brutally went over all the things I had done wrong and areas that needed improvement. I barely heard him say, "But you are safe for solo." On Thursday, I floated out to the flight line and climbed into my trusty T-34B Mentor. I pretended that the instructor was still in the back and made all the required reports to myself over the intercom system. I flew to an outlying field, did a couple of touch-and-go's, and then returned to Saufley because I did not want to do anything that would screw up my chances of being commissioned on Friday, the next day. I was the only member of my class to solo before commissioning, but I always remembered those folks that had not flown that week. I was commissioned on January 18, 1974.

My grades in flight training were below average in fixed-wing and a little above average in helicopters. I was not a natural pilot; I was one of those folks that needed multiple repetitions to learn something. The more I did it, the better I got at it. But I am sure it was a painful process for the folks in the other seat. I do not think you should be an instructor in the training command unless you have had a down or what came to be called a signal of difficulty. That meant you flunked that hop, would get two additional flights to correct your deficiencies, and if you passed the recheck, continue with flight training. Instructor Pilots who have not ever been given a down cannot imagine what a crushing feeling that is. I took two downs going through the fixed-wing syllabus, both while flying the T-28 Trojan in VT-6.

Following completion of initial flight training at Saufley Field, the Navy would determine your pipeline, whether you would fly jets, props, or helos. Jet guys were the pick of the litter and went to Beeville or Kingsville, Texas, or stayed in Pensacola. Prop guys went to Corpus Christi, Texas. Helo guys went to Whiting Field in Milton, Florida, outside of Pensacola. The determination was made based on your grades in initial flight training and the needs of the Navy. My grades and the needs of the Navy said helo's, so it was off to VT-6 at Whiting Field.

The T-28B/C Trojan, also known as the "Air Hog," was the advanced fixed-wing trainer for helicopter pilots. It was a big step up from the T-34B in size and power. You had to re-learn some of the techniques that you developed while flying the T-34, or as we called it, the Teeny Weeny. You could torque-roll a T-28 if you came on with too much power too quickly. The "gouge" was that if you did not see oil leaking from the aircraft, then it needed more oil because it had run out. You could sit in the engine run-up area with the canopy open, put your arms on the canopy rail and look down on the T-34B students. When you ran that Wright R-1820 engine up to full power, you knew you had a lot of airplane on your hands, especially at that stage of our training.

My learning curve with the T-28 was steep, especially because of my monkey see – monkey do way of learning. My first spin in the T-28 was almost my last. The T-34 naturally wanted to come out of a spin. The T-28 wanted to bury itself in the ground. The T-34 spin recovery procedures called for you to momentarily apply the rudder opposite to the direction of the spin, neutralize the controls, and then level the wings and pull out. You were taught that, worst case, if you just let go of the controls, the T-34 would return to balanced flight. For the T-28, the recovery was similar; you applied the rudder opposite to the spin but held the rudder full opposite until the spin started to slow. Then you would neutralize the controls.

Prior to my first T-28 spin hop, I correctly recited all the spin procedures to my instructor during the brief and repeated them to him while we were doing clearing turns in the air. I got the airplane to stall and then applied the T-34 spin recovery procedures rather than the T-28 spin recovery procedures. As I applied the wrong procedures, the aircraft paused momentarily, shuddered, and then entered the most violent spin I had ever, or would ever, experience. After about three revolutions and my trying desperately to figure out what had gone wrong, the voice of God came over the ICS and said, "I've got it." He then said, "you know, Ensign Coyne, I'm not sure if I wasn't here that you would have recovered from that spin."

I promptly replied, "sir, if you hadn't been here, I would be a smokin' hole in the ground!" We went over what I had done wrong, and I spent the next half hour doing spins and spin recovery until I had them perfect.

I received back-to-back downs during the basic familiarization and instrument familiarization portions of the syllable. The first one was for bad basic air work because I was behind the aircraft, and the instructor said I just needed more practice, and the two extra hops would get me where I needed to be. The extra hops would be with one of the standardization pilots in the squadron, in my case, a Marine Corps Captain. The next two hops, the Captain turned me every way but loose. We did all the normal and emergency procedures and covered aircraft systems and limitations, all while flying. My basic air work and confidence increased to the point where he said I should be continued in stage. I promptly went out and got another down on my next flight, which was an instrument flight. I had gotten behind the aircraft and given up.

Because of the two downs, I had to go to a Student Pilot Disposition Board (SPDB) or a Speedy Board. The board would decide whether I would be allowed to continue with flight training or be washed out of the program. I doubled down on my studies in preparation for the board and reported as ordered. The board went on for a couple of hours, and the instructors raked me over the coals. There was bad cop, worse cop, and worser cop. Finally, worser cop looked at me and said, "why the hell are you in here? You know the aircraft, normal procedures, emergency procedures, and instrument flight rules. What is your problem?" I said I really did not know, which I did not, but that I wanted to stay in the program and would do whatever it took to do so. I left the room, and a brief time later, they called me back in and said they were recommending two extra hops, a re-fly of the instrument hop, and if I completed them successfully, then I would be able to continue in the program. But worser cop told me that this was my last chance, and any problems would result in my being washed out of the program.

I went to the flight scheduler to tell him what the board had recommended so that he could schedule me for the practice hops. He asked, "how did the Speedy Board go?" I told him I would rather have a root canal without any anesthetic than go through that again and that Worser Cop was clearly out to get me. He looked at the schedule board and said, "you went before the wrong Speedy Board. That board was put together for another pilot, who needed a kick in the ass. You were supposed to have the kinder, gentler Speedy Board." He scheduled me for the two extra hops that I aced, and then I was scheduled to repeat the instrument flight that I had failed.

When I called in to see who the instructor would be for the flight, the scheduler said that was still up in the air. When I reported the next evening for the instrument flight, in walked Worser Cop. He was all business and said just fly the procedures required for the hop. I was in the back seat of the aircraft and would be under the bag for the flight. That meant that there was a cover that you put over the rear cockpit so that you could not see anything outside and had to depend totally on your instruments. To navigate, you tuned in to local navigation aids, which could also be used to listen to commercial radio stations. Several times during the flight, the instructor switched the radio to a commercial radio station to listen to music. After about an hour of the flight, having done all sorts of instrument maneuvers, instrument approaches, and emergency procedures, Worser Cop told me that he had control of the aircraft and to take a break. Then he told me I could take the bag down. As I was doing so, he rolled the aircraft inverted, and I was now looking "up" at the Florida countryside. In the debrief, he said I flew the aircraft extremely well, and he could not figure out why I had taken back-to-back downs. I nodded my head silently while thinking, that makes two of us, but it will not happen again, and it did not.

The fixed-wing training in the T-28 was at North Whiting Field. Initial and advanced helicopter training would take place at South Whiting Field in the TH-57 Jet Ranger and H-1 Huey. I remember walking into HT-8 for initial helo training, and there was a whole different vibe to the place. It seemed much more relaxed. During our Indoctrination, the Commanding Officer came to address us and said, "helicopters are different." He had a cup of scalding coffee on the table in front of him. He made a fist with his right hand and placed it on the desk. Then he took the cup of hot coffee and placed it on top of his fist. Then he moved his fist gently in a circle so that none of the scalding coffee would spill. "Gentlemen, this is how you fly a helicopter. Your hand movements must be extremely smooth. You don't want to wrestle the aircraft around the sky. You want to caress it!"

You never forget your first flight in a helicopter. We flew a lot lower than we did in fixed-wing aircraft. We also flew a lot slower, which meant that you could see a lot more, and more importantly, you could get back quicker to those things that you wanted to see a lot more of. The instructor demonstrated all the capabilities of the helo. We moved forward, then we moved to the side, stopped, and sat there in a hover. We went up and down. I thought to myself, this is cool. Then we backed up, and that sealed the deal. No more regrets about not going jets.

Now it was my turn to take control of the aircraft. The instructor first tells you to take control of the rudder pedals. The pedals change the pitch or angle of attack on the tail rotor, which causes the aircraft to turn right or left in level flight. The initial tendency is to "walk on" the rudder pedals, and so the aircraft moves a little left and then a little to the right, with the excursions from straight ahead getting less and less as you get the feel of the maneuver. Next, the instructor gives you the collective. The collective controls the pitch or angle of attack on the main rotor blades. Pull up, and you begin to climb or go up, lower it, and you descend or go down. The other thing is that as you increase or decrease collective, you must make an adjustment on your rudder pedals. You step on the left pedal when you increase power/collective and the right pedal when you decrease power or lower the collective. This is exactly the opposite of the rudder control inputs you use in a fixed-wing aircraft, so on your initial flight, you must consciously think about which pedal to step on when you make a power adjustment. But hey, you have two hundred hours of flight time, so you adjust quickly. Your control inputs for the rudder pedals and the collective get better and better. Now comes the hardest part – the cyclic. The cyclic controls the pitch or angle of attack on the main rotor blades and moves you left, right, forward, or backward. The instructor has the helo in a perfect hover and tells you to get on the controls with him. He gives you a few seconds and then says, "You have the aircraft."

What follows is like being at the rodeo when they open the gate and the bronco rider comes out. The aircraft looks like a bucking bronco as it moves in every direction simultaneously. Before things get too out of hand, the instructor takes it back, steadies the aircraft, moves it back over the landing pad, and once again says, "You have the aircraft." The second time is not much better than the first, but the excursions in the student pilot's control inputs get to be less and less. In the debrief of that initial hop, the instructor usually tells the student to take a long hot bath that evening because of all the muscles that you strained on that first hop. I took the advice and the bath but was still sore as hell the next day.

Amusement parks make a lot of money simulating the things that you can do in an aircraft. Things like loops and barrel rolls. The e-ticket ride in a helo is doing autorotations. There is a school of thought about helicopters that says if something bad has not happened, it is about to! About the worst thing that can happen in a helo is that the engine or engines fail. While highly unlikely in fleet multi-engine helos, the Training Command's Jet Rangers and Hueys only have one engine. You are now possibly in the worst-designed glider in the world. You will glide as far as you can see or fall, which is about straight down. You are trained to pick the best alternative of the worst available landing fields that are under the aircraft and then try to land there.

The first phrase that every helo pilot learns is "down, right, turn." These are the immediate action items for an autorotation. Why would you have to autorotate? Things are not going well in the aircraft. Like engine failure, main gearbox problems, electrical fire, smoke in the cockpit, etc. In short, anything that requires you to be on the ground immediately.

"Down" refers to what you are going to do with the collective. You immediately push it down to the deck (floor). This will cause the aircraft to descend immediately. When you lower the collective completely, the rotor system is unloaded, and the inertia on the main rotor blades will increase as you descend, allowing the aircraft to build up energy. That energy will be necessary to turn into lift, or cushion when you arrive at the bottom of this wild ride.

"Right" reminds you to put in the right rudder to counterbalance the loss of power that will cause the aircraft to want to rotate to the left. In a helo, you are constantly using the rudders to keep the aircraft in a balanced flight. If you are not in a balanced flight, you fall faster, which reduces the amount of time and distance that you have while looking for a place to land.

"Turn" towards the place where you want to land. When you start flying helos, you are taught to fly from landing spot to landing spot. If you must autorotate right now, what field are you going to land in? After a while, this becomes second nature, and you do it by instinct. You learn to do straight in autos, and the field is right in front of the nose of the aircraft. Ninety-degree autos involve a 90-degree change in course to get into the field. The most challenging autos are normally 180 degrees to get into a field that you just passed over and is now behind you. They involve the most maneuvering, are the most fun, and result in the most beers won or lost in the club that week at happy hour. The practice autorotation criteria are normally an autorotation to the numbers on the runway or even to the landing pad at an outlying field. The pilot closest to the intended point of landing wins. Over time you get good at judging what field you can make it into and the different kinds of autorotation necessary to make it into the chosen field. The sadists do 360-degree autos. Why would you do that? Because the best field is right under you, and you need to lose altitude in order to be able to make it into that field.

Unlike in fixed-wing training, I did not have any difficulties in either primary helicopter training in HT-8 or advanced helicopter training in HT-18, unless you count the crash in Alabama. I enjoyed flying and began to get more comfortable in the air. The amount of time I spent in the pool waiting to go from HT-8 to HT-18 was minimal. I do remember reporting to HT-18 and asking a fellow student if we had been assigned a flight leader. He said we had and gave me the name. Luckily, the high school I attended had a lot of students of Polish descent, so the name did not throw me off. But I was dismayed to find out that he was a Marine. The Marines, whom I had a lot of experience with, remember my fixed-wing adventures, tended to be sticklers on everything from preflight to normal procedures. I asked what our new flight leader looked like. My fellow student said you would know him when you see him. I asked how I would know it was him when I had never met the man. He said, "he blots out the sun." Later that day, I was walking down the passageway (hallway) when the sun suddenly disappeared, and I knew I had found Captain "Z." The guy was great, and I could not have had a better flight leader in the Huey.

As time went on, the prospect of getting our wings of gold became more likely, but you never wanted to take anything for granted. Your flight grades had determined that you were assigned to the helo pipeline, and now your grades in the helo pipeline were going to determine where your first duty station was going to be. The top student in the class was given their choice of orders. The rest of us took what was left. Everyone wanted to go to the fleet in a helicopter squadron. At that time, the squadrons were HS (Helicopter Anti-Submarine), HSL (Helicopter Anti-Submarine Light), HC (Helicopter Combat Support), and HM (Helicopter Mine Sweeping). If you did not get a fleet seat, then you were starting behind the power curve career-wise. The week I came up for orders, the top guy got a fleet squadron, and the rest of us, some twenty-plus folks, were assigned to Naval Air Stations. I went home to tell my wife and found out that one of the other student wives had already called and told her we were going to Bermuda.

My first flight in the T-34B took place on December 6, 1973; I got my wings on December 13, 1974. It took me just over a year to go from Aviation Officer Candidate to Naval Aviator. I was about to become a nugget, an Ensign in his first flying operational assignment.

CHAPTER IV

NAVAL AIR STATION BERMUDA

JANUARY 1974 – OCTOBER 1977

I know that I should have been thrilled at being assigned to Naval Air Station Bermuda, but I was not. I did not consider all the unique professional opportunities and individual experiences that would come with living on the island. I was thinking only about the fact I was not going to the fleet and that I would be behind all my contemporaries because I had not done so. I was a reservist, and if you wanted any career in the Navy, you had to be a regular. You got to be a regular by being in a Fleet Squadron, not by serving at an Air Station. Only in retrospect, years later, did I realize that it was exactly where I needed to be at that point in my naval career and that the things I learned at the Air Station would serve me well throughout the rest of my career in the Navy.

My first job was as the Avionics Division Officer in the Aircraft Intermediate Maintenance Department (AIMD). I did not know it at the time, but being involved in maintenance is where I would establish my reputation as an aviator. As the Avionics Division Officer, I had two Chief Petty Officers, AEC English and AEC Gunning, and twenty enlisted folks working for me.

As a division officer, I was responsible for the repair of all the avionics equipment associated with the aircraft based at the Naval Air Station. We supported the C-131 Samaritan base logistics aircraft, the deployed Anti-Submarine Patrol (VP) P-3 Squadron, occasional detachments of Fleet Air Reconnaissance Squadron Four

(VQ-4) C-130s, and our station UH-1N Hueys. I learned the meaning of the Navy Enlisted Classification (NEC) codes and what pieces of avionics were associated with which aircraft platforms. I ensured that my folks were professionally trained, and I was responsible for ensuring that they had the right schools and equipment to do their jobs. I was also responsible for scheduling their work hours and liberty and making recommendations for advancement exams, awards, billeting, etc.

My sailors worked out of a couple of trailers that were located outside the AIMD building. My office was on the second deck inside the AIMD building. Approximately six of the folks that were working for me came from the deployed VP squadron who sent them to us on Temporary Additional Duty (TAD). Managing these folks was a challenge as they were usually the sailors that the VP squadron felt that they could spare, not necessarily their top performers. In our pre-deployment meetings with the VP folks, we would identify the areas and pieces of equipment on which we lacked expertise and request that the folks they sent TDY to us had recent experience working with those pieces of equipment. Often, the TAD VP folks had been to the required school and had worked on the gear, but that was a number of years ago. The idea that the squadron was hurting itself by sending sailors to AIMD who could not fix the gear that the squadron really needed to be fixed never seemed to enter their thoughts. I also was amazed that fully mission-capable (FMC) aircraft could leave Brunswick, Maine, or Jacksonville, Florida, for Bermuda and suddenly have an overwhelming number of discrepancies immediately after they landed in Bermuda. The fact that most of the broken equipment was directly involved in Anti-Submarine Warfare made me wonder if they had even turned it on until they were on final approach into Bermuda. So, we would work twenty-four-hour days for as long as it took to get the equipment fixed and back to the squadron. Then we were able to assume a more manageable work schedule.

My two Chief Petty Officers did a fantastic job of shaping me as a division officer. I remember one day griping about not being in the fleet, and one of them replied, "the Fleet isn't all it is cracked up to be." I learned about the people in the division, their backgrounds, and aspirations, and I tried to figure out how I could support them. Thanks to my sailors, the avionics division ran very smoothly. That allowed me to devote a lot of time and energy to learning how to fly the "Twin Huey." I had flown the single-engine variant in the training command in HT-18, and I was familiar with the basics of how to fly the aircraft. Obviously, having a second engine was a comforting thought. The second engine gave you more power than you would ever need for the Search and Rescue mission.

At that time, there was no Fleet Replacement Squadron (FRS) to get training in the twin-engine Huey prior to reporting to Bermuda. All that training would take place on the job or OJT. The helo pilots were a mixed bag, as you might expect. The senior helo pilot was a Commander and the Safety Officer for the Air Station. He had begun his career in fixed-wing aircraft prior to transitioning to helos. The other senior pilots were Lieutenants. Their backgrounds included a composite squadron (VC), Helicopter Attack Light Squadron (HAL-3), and Helicopter Combat Support Squadron (HC-7). The last two had Vietnam experience, including a Navy Cross for making a rescue of a pilot under fire in North Vietnam. Two other first-tour pilots had been there for two years, and two pilots were completing their tours, one of whom I would be replacing. I tried to learn everything I could from them. I quickly fell into a pattern of flying once or twice a week, a couple of weeks a month. Our primary mission was Search and Rescue (SAR). We are available to support the operations of the VP and VQ folks as well as the civilian commercial air traffic. NAS Bermuda was a joint military, civilian airfield with the US Navy on one side and commercial aviation on the other. We also were tasked with providing SAR support for the trans-Atlantic flights of tactical jets that would occasionally come through.

We practiced our mission daily, and I developed the ability to hold the Huey in a steady hover. We would practice the deployment and recovery of our SAR Swimmers. The deployment consisted of a jump into the water at ten feet and ten knots. If you established those parameters, then the swimmer would enter the water easily. If you were higher or faster, then the swimmer would skip across the water. I gained a greater appreciation for those parameters when I jumped from the aircraft to provide a downed pilot for the crewmen to practice on. I had my fair share of skips across the water. Those experiences made me get even lower and slower when I would put a crewman into the water.

The SAR practice was broken up by real-world missions that involved picking sailors up off underway submarines, medical evacuations from other US Navy ships, and providing support to Argus Tower. Argus Tower was an acoustic research tower that was about thirty miles southwest of the island, and the landing platform was one hundred feet off the water. The platform was large to land on, but you could only shoot the approach from certain directions, and that sometimes made it a bit tricky. We were also called upon to supply a jump platform for the United Kingdom (UK) Royal Marines to parachute out of during the Queen's Jubilee Celebration. If you have ever seen the motion picture "The Deep" with Nick Nolte and Jacqueline Bisset, it opens with a scene where the camera descends out of the clouds around Bermuda and then under the water to some skin divers. I was one of the pilots that flew the film crew during that scene. The film crew presented us with some gifts. I treasured "The Deep" tee shirt for a few years but made short work of Gosling's Black Seal Rum.

My two most unique flights in Bermuda involved flying while inebriated and saving a life.

It was over the Christmas holiday, and I was on SAR duty as the copilot. There were only three helo pilots on the island, and we had flown a hop earlier in the day. LT Frank Staebler was hosting us for dinner, and the third pilot (LTJG Art Gallo) was also there. We checked in with the Operations Duty Officer to see if the VP or VQ folks were scheduled to fly that night. The answer was no, so we lit the drinking lamp. We had cocktails and wine with dinner and were sipping our after-dinner drinks when the phone rang. As I remember the conversation through my alcohol-soaked brain, it went something like this. Frank listened to the caller on the other end, and it seemed the first words out of his mouth were, "you want us to do what? We have been drinking. No, he is here (Art Gallo), and he has been drinking. Are you sure that you want us to launch? OK, we'll be there shortly, have them get the aircraft ready." He hung up and said that the cargo ship that brought supplies for the Navy in Bermuda had run aground and was sinking. They wanted us to launch and pick up any survivors.

We arrived at the air station and began pre-flighting the aircraft. The crewmen showed up and said they had been pulled from the club for the mission and had been drinking. That made four of us. When we were about to start the aircraft, the CO of the Air Station showed up and said he wanted to go on the flight. It was immediately apparent that he had also been drinking. So, we all piled into the aircraft and departed for the sinking ship. When we got to the site, the ship did not appear to be sinking.

Frank pulled into a hover and manipulated the searchlight to give us a complete view of the ship. Our Huey did not have a doppler or coupled flight system to assist in holding a steady hover. The aircraft would only stay where you physically put it. After establishing the initial hover, Frank began a very slight increase on the collective, which caused us to gain altitude. This was a response to being slightly inebriated on a black-assed night with no horizon in an aircraft full of slightly inebriated individuals.

The Base CO was on a headset, so I did not want to say anything on the intercom system (ICS) to Frank, and I gently pushed the collective down. Frank interpreted what I had done as the aircraft descending on its own and increased the collective even more. I tried another gentle push with an even greater reaction from Frank. We were still climbing when I came on the ICS and said, "altitude." Frank looked over at me, shook his head, and we worked our way back down to one hundred feet. It became clear that the ship was not going to sink, and there were no survivors to rescue. We figured we had pushed our luck about enough, so the inebriated crew in the airplane decided to return to the air station. After we shut down, the CO thanked us for our efforts and left. We all looked at one another and said, "I'll never do that again!" and I never did.

The saving of a life involved a stokes litter pickup off a US Navy Gearing Fram II destroyer in the middle of a thunderstorm.

We got the call that a sailor on a US Navy destroyer had a life-threatening condition that required an immediate Medical Evacuation or MEDEVAC. As I recall, the ship was estimated to be at the limit of our available fuel onboard the aircraft. It had a flight deck for drone operations, but we would not be able to land. Given the sailor's reported medical condition, I intended to use the stokes litter rather than the rescue hoist to take him off the ship. I knew that we would not have a lot of time on station due to the winds and nasty weather being generated by a front in the area. We found the ship in an area that had storms all around it.

I pulled into a hover, and we lowered our second crewman. We had used about half our gas to get to the ship, but I thought the winds would be favorable for our return. Nevertheless, I instructed the crewmen that we wanted to do this as quickly as possible. The first crewman wanted to move off the ship to let the second crewman work, but I was concerned about losing the ship in the weather. Also, the first rule in SAR was never to lose sight of the survivor. The second aircrewman went into the hangar to assess the sailor's condition. He gave us a thumbs up, and we lowered the stokes litter. The stokes litter was a stretcher with wire mesh walls to keep the patient from rolling off the litter. The second crewman strapped the sailor into the litter and took hold of the guideline that would keep the litter from spinning.

The ship was moving quite a bit, and it took all my skill to try and keep us stationary over the spot as the litter came up. It seemed like a long time, but the litter got to the doorway, and the first crewman swung the hoist into the aircraft and detached the litter. The hook immediately went back down, and the second crewman was in the aircraft within a couple of minutes. We closed the door and made the best speed for Bermuda. Due to the patient's condition, they wanted us to get him to immediate medical attention at the Air Station rather than take him directly to King Edward VII Hospital. Later that day, I heard that he made it but that if we had not gotten there when we did, he would have died.

A couple of weeks later, the sailor had recuperated enough to go back to the states for treatment. My wife worked at US Customs in Bermuda and overheard this gent telling the customs folks about how he came to be on the island without a passport or any papers. He said he had been extremely sick on the ship and had been told there would be an attempt to get him a MEDEVAC. They moved him into the DASH (Drone Anti-Submarine Helicopter) hangar and said the helo was coming to take him to Bermuda. He knew the weather was bad. There was lightning, thunder, rain squalls, and high winds. After a while, they told him they did not believe the helo would be able to make it to the ship. He began to doubt that he was going to make it. Then he said he heard the most beautiful sound in the world. The sound of the Huey blades slapping the air in that distinct whop, whop, whop. Then he knew he was going to make it. It brought tears to my eyes the day Linda told me about seeing him. It brings tears to my eyes today thinking about it.

Then there was the day I took the helo to the beach.

We had completed our assigned mission for the day and were on our way back to the air station when the motion on the instrument panel caught my eye. I thought I saw the temperature/pressure gage for the main transmission fluctuate momentarily. I watched it for the next few minutes, and nothing happened. As I began to relax, I saw it fluctuate again. I confirmed with the copilot that he had seen the same thing. We were transiting along the south shore about a mile off the coast, and I decided to get closer to land in case we had to put the aircraft down. A few seconds later, the main gearbox pressure began to fall slowly. The aircrewman looked out along the side of the aircraft but said he did not see any transmission fluid on the fuselage. Although I did not have any chip light or other secondary indications, I decided to land on the first flat piece of ground that I saw, which was the beach. I broadcast a mayday, and we made an uneventful landing.

As a naval aviator, you would think I would be cognizant of a little thing called the tides. I landed at low tide, but I knew that in a matter of hours, this piece of the beach would be underwater. The air station sent some SEABEES and security folks to help us figure out what we would do next. We cleaned the chip detector, which did not have anything on it, and inspected the aircraft for leaks. There were no leaks, and the transmission oil reservoir was full. The

SEABEES put some planking alongside the helo to keep it steady when we started the engines and engaged the main rotor. One of the senior helo pilots had arrived and would take over as the aircraft commander. Our plan was to turn the aircraft up and see if we got the proper temperature and pressure. If that worked, we would pull into a hover for a few minutes. If all went well, we would proceed at low altitude and airspeed back to the air station. All the checks went OK, and we were soon back at the air station.

Everyone was looking at me like I was nuts because we could not duplicate the main gearbox indications. After a more detailed inspection, we took the aircraft out on the flight line the next day to evaluate it again. We sat there turning for about 45 minutes and could not duplicate what had happened to me in the air. AD-1 Lester Rice was a senior first-class petty officer who had started out in the Navy repairing reciprocal engines and had transitioned to jet engines. He was a good old boy and had me go over everything we had done on the flight several times. He was kneeling behind me and suddenly reached forward and hit the instrument panel where the main gearbox pressure and temperature gage was located. The pressure and temperature started to fall. It momentarily stabilized and then returned to normal. He hit the panel again with the same results. He said, "Lieutenant, you ain't crazy; we just got a bad electrical connection. Let me get an electrician out here, and we'll change it out and see what happens." We never had another problem with that aircraft. I had to answer a bunch of questions from the safety folks about what I did and why I did it, the foremost being, "why didn't you go another two hundred yards and land on the fairway at the golf course?" I felt like I had screwed up when one of the experienced Vietnam pilots said, "if you are here answering their questions, then you did something right, especially compared to the alternative."

My growth as a Naval Officer continued with a variety of assignments while stationed in Bermuda. I spent six months as the Avionics Division Officer. Then I was assigned as the Assistant Communications Officer. The Communications department managed all the message traffic from the air station to the world. Radiomen always knew what was going on because they were the ones who processed the messages and routed them to the applicable departments. A radioman could not help but scan the messages and therefore was at the heart of RUMINT, or rumor intelligence. They knew before anyone else in the organization what was going to happen next. I was going to work for a Lieutenant and would have another Ensign, a Master Chief, two Senior Chiefs, and about one hundred folks working for me. My appreciation for Chiefs running the Navy took on a whole new meaning when I went to Communications.

The Chiefs regularly invited me to their Club on Fridays for extra instruction. They also appointed me a defense attorney for a newly selected chief. My father had alluded to the CPO initiation ceremony but did not go into a lot of detail. The fact that I was invited, and as a defense attorney to boot, meant that I had a good relationship with my chiefs. In my duties as defense counsel, I required quite a bit of truth serum, a concoction of various liquors and things you did not want to know about. After several rounds of truth serum, the judge asked me, "how does your client plead to the charges?" which included masquerading as a Chief Petty Officer. I replied, "not guilty by reason of insanity, your honor!" The judge was incredulous and asked, "did you just say not guilty by reason of insanity?" I replied, "yes, your honor, he would have to be insane to want to be a Chief Petty Officer." The rest is an alcoholic haze.

One day my Department Head was flying the base bird to the states when the captain's office called over and said the CO wanted to see him. The yeoman told the CO's secretary, who really ran the base, that the Department Head was not available, and the secretary replied, "then get the assistant department head over here immediately!" The yeoman came into my office with Master Chief Neville in trail and said I needed to get over to the CO's office most ricky-ticky. I looked at the Master Chief and asked him, "what do you think this is about?' He said he did not have a clue. I then went through a list of projects we were working on and asked, "do you think it is any of them?" The Master Chief, who had never been on shore duty prior to reporting to Bermuda, paused for a moment and said, "Lieutenant, you go over there and tell the captain anything you want, then come back here, and we will make it so!"

My next assignment was the closest I ever came to getting out of the Navy.

I was happy in Communications and not looking to move when I heard a rumor that I was being considered to run the Officer's Club. The Officer's Club was at the hub of social activities at the air station. This was back in the hard-drinking, hard-smoking, woman-chasing, look the other way, and politically incorrect Navy. Every Friday, there was usually an officer's call that required mandatory attendance. There were wetting downs, where an officer who had been promoted would host his fellow officers and their spouses to drinks and food to celebrate his good fortune. On the holidays, there were always parties at the club. The food and drink were cheap, especially at Bermuda prices, and you could play the slot machines if you were not playing a dice game or shuffleboard. I would usually arrive at this event having spent the afternoon in the CPO club. Other than being Irish and liking to drink, I could not figure out why the CO would consider me to run the club. The real secret to my selection is that all the folk's senior to me were smart enough to produce some excuse about why he should not select them. The CO had run an Officer's Club in Pensacola when he was a junior officer and believed it was one of the reasons he had succeeded in naval aviation. That should have been my first indicator that this would not turn out well. When the CO had run the Officer's Club in Pensacola, he had military personnel assigned to perform the work required in the club, essentially free labor. Everyone that would be working for me was a civilian that had to be paid full salary plus benefits. There was no way I would be able to provide the same sorts of services for the same price that the CO had been able to do in Pensacola. We were on a collision course.

I was smart enough to know that I did not know anything about club operations but that the Navy had a school, Mess Management School. I know you are laughing, but the school at Patuxent River, Maryland, taught you how to manage a Commissioned Officers Mess Open (Ashore) (O Club), consolidated package store (liquor), and Officer and Enlisted Bachelor Quarters. There were a lot of ways to get in trouble running these operations, and the course was designed to keep you from making bad decisions, or at a minimum, to know that you were making an unwise decision and one that could get you sent to jail. I graduated first in my class. I learned three things at Mess Management School: you cannot lose money on a bar, you cannot make money on a food operation, and it is just a matter of time before you are fired. I could hardly wait to return to Bermuda to assume my duties.

I tried my best to improve the operations at the club. We produced new menus, different entrée's, publicity, and live entertainment. Nothing I tried seemed to please the CO, and the critiques would take place in the club on Friday nights in front of my superiors and subordinates. The redeeming quality was the people that worked for me. They were great. I remember in school; they told us to pay attention to the books and let our accountant know that we were watching. I remember walking into the accountant's office one day and saying, "Graham, how do I know you aren't siphoning money from our operations?" A Bermudian national civilian employee, he laughed and said, "Dear boy, you don't make enough money on this operation to even begin to tempt me to do so." He had been at the club long enough to see about a dozen or more folks go through my position. He was the guy who saved me from going insane.

I did not know what a panic attack was at that time. I began to think it was normal to wake up at two in the morning with my mind racing on all the things I had to do the next day and trying to determine what the CO would be angry about next so that I could try to mitigate the damage. The thing that really upset me was that it was beginning to cut into my flying. The CO would call the club with a question. When they told him I was out flying, base operations would call on the radio and tell the aircraft to return and have LTJG Coyne report to the CO's office. Every hop was, "Will he or won't he call me back today?" I was extremely proud when I left the club after about a year that we managed to turn a slight profit. Did I mention that it is hard to lose money in a club when you can serve alcohol and have slot machines?

I was banished to Operations as the Fire Division Officer with about twenty Aviation Boatswains Mates working for me in the Fire House. We provided crash and rescue services for the military and civilian operations at the airfield. I would sit in the fire truck on the side of the runway to get to know my folks. I dragged hose, put out fires, and went to Captain's Mast almost every week. Aviation Boatswains Mates are the salt of the earth. They live for the moment and enjoy taking risks when consuming alcohol, which often results in them being on the wrong side of the law. When they do, they go to Captain's Mast. In Bermuda, when a sailor went to Captain's Mast, also known as an Article 15, the CO required the Department Head, Division Officer, Division Chief Petty Officer or Leading Petty Officer, and the accused (also known as the guilty bastard) to be in a service dress uniform. There were a couple of reasons for this. It confirmed the solemnity of the occasion and was a pain in the ass for everyone, thereby hopefully motivating all concerned not to do it again. This might have applied to normal folks, but not my guys/gals. I was there to stand up for my people as they tried to explain why they had done things like break into their post office box because they forgot their combination or mistakenly took someone else's vehicle because they could not find theirs. Of course, alcohol contributed to their inability to make good decisions. You get the gist. Work hard, party harder, and then accept the consequences.

A properly held Captain's Mast has many things in common with a good Broadway play. You have the protagonist (CO), antagonist (guilty bastard/accused), and witnesses (Department Head, Division Officer, Division Chief Petty Officer = audience). The CO calls everyone but the accused into his office and goes over the circumstances surrounding the offense. He wants to know the background of the accused, what "allegedly" happened, an assessment of the future potential of the accused, and recommendations for sentencing. There are several outcomes of the Mast depending on the recommendation of the Department Head, Division Officer, and Chief Petty Officer. The accused can be found guilty, and the range of punishment falls into the following categories: Maximum – Monetary Fine, Reduction in Rank, and confinement. Medium, suspension of all of the above if the accused does not commit another offense within a specified period of time. Minimum, the CO throws out the charges and gives the accused "a good ass chewing."

Depending on the nature of the offense, the recommendations of the chain of command, and the general demeanor of the accused at Mast, COs can adopt a Jekyll or Hyde mentality. My first trip to Mast was unforgettable. The offense was minor, the chain of command favored suspension, and then the accused was brought in. While the CO was questioning the accused, it became apparent that the CO did not think the "guilty bastard" was taking this seriously enough. In a split second, the CO's demeanor went from one of a concerned father trying to understand a miscreant child to a homicidal maniac that I thought was going to jump over the podium and strangle the accused. The CO found him guilty and threw the book at him in terms of maximum fine, reduction in rank, and confinement. When he finished the sentencing, the CO thundered at the Master at Arms to "get this piece of shit out of my sight." When the door closed, the CO turned to us and as calmly as could be, asked, "what is the story of our next defendant?"

My other experience with the military legal system was serving on several Courts-martial while in Bermuda. These were more serious than a Captain's Mast, with the possibility of severe punishment and consequences, including time in the brig and discharge from the service with a bad conduct discharge. The technical term for this was a discharge under other than honorable (OTH) conditions. An OTH meant no educational GI Benefits and no medical care, and your employer would inevitably want to know why you got kicked out of the service. Not a good thing when "The Greatest Generation" of World War II veterans were the ones that would be deciding on whether or not to hire you.

For most Courts-martial, a defense attorney and a prosecuting attorney would fly in from the states. It is hard for the members of the board to keep their preconceived notions from affecting their judgment. I would inevitably inspect the Judge Advocate General (JAG) Navy lawyers to see how they wore their uniforms. It was that Marine AOCS indoctrination coming back. If the lawyer did not have his ribbons on correctly, a good gig line (line of the shirt, line of the trouser fly opening, and the front edge of the buckle are not all lined up), properly shined shoes, and a decent haircut, then they lost a little credibility with me. Sometimes the prosecutor would be the Air Station JAG, and he enjoyed home-field advantage because we ate lunch and played foosball with him every day. I learned that Courts-martials should be used for only the most serious offenses. Once we were sitting on a Courts-martial where it was obvious that the accused was not guilty and that the case should have been managed at Captains Mast. The defense argued that the accused was a choir boy, and the prosecutor made him sound like a mass murderer. Our deliberation was quick, and we returned a verdict of not guilty. After the trial, the attorneys on both sides said we had made the right decision. I thought to myself, if you thought he was innocent, then why did we have this trial? If they felt that way, why did they press the charges? If the members of the court were having a difficult day, then this kid might have ended up in the brig.

Over the first two years in Bermuda, I thought about what I wanted to do for my next assignment. A detailer is the officer's representative who works with you in figuring out what your next assignment is going to be. The placement officer represents organizations/squadrons that require personnel. The dialogue between the two has a lot to do with a Naval Officer's career. I had been encouraged to establish a relationship with the detailer. The idea was that the more information the detailer had about you and what you wanted to do in the future, the better he could match them up with vacant positions. I also wanted to believe that it is harder to send someone to a bad job when you have looked them in the eye and established a personal relationship. I went all in and took Linda to see the detailer on one of the periodic visits. This gave the detailer a better idea of who I was as a person, and he did not have to rely solely on my fitness reports to figure out where I should go next. That personal relationship is what saved my naval career.

My first tour was shore duty in Bermuda. My next tour should be on sea duty in the fleet. Coming from an air station meant that I had not made the cut out of the training command to get a fleet seat, or an operational flying billet, in a squadron. Any hope of my having a naval career required augmentation into the regular Navy, and that meant a squadron assignment. An Officer's Fitness Report (FITREP) was used to determine your competitiveness for future assignments. In order to be competitive for a squadron assignment, you needed good FITREPs. FITREPs were informally divided into categories called Pack Plus, Pack, and Pack Minus. An officer was expected to move from the lowest (pack minus) to the highest (Pack Plus) through the course of their tour in a given assignment. If you deviated from the progression, there had to be a reason. A declining fitness report, going from pack-to-pack minus, indicated the officer was not doing well and would not be considered for challenging future assignments.

Pack was a normal naval officer. If you started in the pack plus category, then you were doing well and would be singled out for the best future assignments. If you were consistently in the pack minus category, then it was time to consider a career outside the US Navy. The detailer stressed to me that I needed a good fitness report coming out of Bermuda to be competitive for a fleet assignment.

I wanted to go to the Light Airborne Multi-Purpose System (LAMPS) community and fly H-2s from frigates and destroyers. LAMPS folks deployed in detachments rather than as a whole squadron. You could be an officer-in-charge (OINC) as a Lieutenant and would go to sea on a frigate, destroyer, or cruiser with fifteen folks, including pilots, a crew leader, and the various enlisted aviation ratings required to maintain and operate the aircraft. As OINC on a surface ship, you were the Carrier Air Group Commander (CAG), Air Operations Officer (Air Boss), Commanding Officer (CO), or Skipper all rolled into one. You are the aviation expert on that surface ship and responsible for all air operations.

During the detailer visit with Linda, he gave us the tour, showed us where the selection boards met, and observed the detailing/placement process in operation. Various officers would wander around with strips of paper that had an officer's history on them. The placement officer would state the qualifications to fill a billet in an organization. The detailer would scan his strips to see if he had an individual who met those qualifications. The detailer and the placement officer would use that strip in determining your future assignment in the Navy. Linda wondered what would happen if the paper strip with your name on it fell behind someone's desk. The detailer stressed I needed a good fitness report to be considered for a fleet seat, trending towards pack plus. I asked about what would happen to me if I did not get a good FITREP. He said I could do an exchange tour in Iran flying the Huey, run the Batchelor Officer's Quarters at some other air station in a non-flying billet, or I could use my radio-television skills at the Navy Liaison Office in Hollywood.

Following the visit to the detailer, it was time for me to get my next FITREP. I reported to the CO's office and saw that my ranking was lower than it was before. The CO asked me what I thought, and I replied, "Captain, I know this FITREP represents what you think of my performance, but it looks like I will be going to another air station and not to the fleet." He said it was a good FITREP and should make me competitive for a fleet seat. I said that was not what my detailer told me. He called his secretary and said to get the detailer on the phone. When the secretary told the CO the detailer was on the line, the CO put him on speaker but did not tell the detailer that he had done so or that I was sitting in the room. The CO told the detailer that he was doing FITREPs and had an especially competitive group and was trying to figure out how to rack and stack them. The CO went on to say that he was considering putting me into a certain category. The detailer replied immediately that a FITREP with those grades would result in my assignment to another air station. The CO thanked the detailer and hung up. I could see that he was changing some of the grades on the FITREP, which I then signed.

The orders came in a few weeks later, and I was assigned to Helicopter Anti-Submarine Squadron Light THIRTY (HSL-30), the Fleet Replacement Squadron (FRS) for the H-2 community on the East Coast. I was assigned to a sea-going detachment in the squadron that provided support to a Naval Research Vessel, USNS HARKNESS TAGS-32, and a deployed Oceanographic Unit. It was sea duty in a shore duty squadron, and in effect, it could provide a back door into an operational HSL squadron. The detailer had balanced my paper record with his personal impressions based on several visits with me. My record did not support going to a fleet squadron, but he felt that I had the potential to be competitive if given the opportunity. By sending me to sea duty in the FRS, I would have the opportunity to prove myself. If I did well enough, it would pave my way into a fleet operational HSL squadron. If not, I would be out of the Navy within two and one-half years.

CHAPTER V

GOING TO THE FLEET – SORT OF?

HELANTISUBRON LIGHT THREE ZERO (HSL-30)

NEPTUNES' HORSEMEN

OCTOBER 1977 – DECEMBER 1979

I Reported to HSL-30 in the fall of 1977. HSL-30 was the Fleet Replacement Squadron or FRS for the H-2 helicopter on the East Coast. Everyone referred to it as the RAG, which stood for Replacement Air Group, a holdover from the earlier days of naval aviation. I would learn all about the H-2 and how to use it tactically for missions involving anti-submarine warfare (ASW), anti-ship surveillance and targeting (ASST), search and rescue (SAR), medical evacuation (MEDEVAC), and logistics support or utility operations (HAULING TRASH). The FRS would teach me how to do all these missions, but my deployment would focus on logistic support and utility operations.

The FRS syllabus was divided into an academic phase where you would learn tactics, techniques, and procedures while "flying" a Weapons System Trainer that replicated the H-2. The second phase involved actually flying the aircraft and learning/practicing normal and emergency procedures. In HSL-30, the tactics and Weapons System Trainer (WST) portion of the syllabus was geographically separated from the rest of the squadron. Once you finished that portion of the syllabus, you then went to the squadron to actually fly the H-2.

My FRS class was unique. There were ten of us, which was a large class. We had a smattering of prior H-2 experienced pilots (4), transitions from other communities (2), Air Stations (2), Training Command (1), and an Ensign Nugget. Three of us would be going to HSL-30 Det Alpha, three to HSL-36 in Mayport, Florida, and four to HSL-34 in Norfolk, Virginia.

Class members were Lieutenant T.C. Jones, Lieutenant Junior Grade Ken Clements, Lieutenant Tom Freeland, Lieutenant Mark Davis, Lieutenant Sandy Clark, Lieutenant Ed Driscoll, Lieutenant Dave Youngblood, Lieutenant Dave Palleson, Ensign Tim Sullivan, and Lieutenant Junior Grade-me. The senior folks in the class adopted a take no prisoners attitude when it came to dealing with the training instructors. They would challenge the instructors when what they were teaching did not match up with their fleet experiences. They won most of the arguments. The training guys would increase the complexity of the tactical problems in the Weapons System Trainer (WST) until you would screw it up. That did not take long for me, but a lot longer for the other guys in the class. The WST had three seats, and in the initial stages, the instructor would ride in a jump seat right behind the pilots so that they could observe and interact with the students more efficiently.

The WST was supposed to fly just like the aircraft, but the consensus among the experienced H-2 guys was that it was harder to fly than the actual aircraft. But they could introduce emergency procedures and tactical situations in the WST that you would never think of doing in the aircraft. You could "crash" the WST, and it was quite a jolt. You knew that terrible things were about to happen when you heard the instructor strapping into his seat. Shortly thereafter, engines would fail, controls would malfunction, and the severity and complexity of the situation would increase to the point at which you would eventually crash. The challenge was to make the instructors work as hard at getting you to crash as you did in preventing the crash.

The instructors would put an emergency of the day on the blackboard in our classroom. The senior guys would erase it and produce their own emergency procedures, including elaborate checklists to be followed. Among my favorites were "snakes in the cockpit" and "tail rotor fire."

I was paired up with the nugget, Ensign Tim Sullivan. We approached the FRS in opposite ways. Tim was incredibly intelligent and an outstanding tactician who could do the tactical problem entirely in his head. He was pretty laisse fair when it came to systems knowledge and emergency procedures. He had distilled the H-2 down to its basics. "Gas goes in, smoke comes out, and rotors go round and round." However, we complemented each other completely as I focused on aircraft systems, limitations, and emergency procedures. When I was running the tactical problem and was stumped, Tim would point to the PT-429 screen to indicate where the next sonobuoy should go. If it were an emergency procedure, I would say, Tim, of course, that is a main gearbox chip light, and as you know, we should immediately land.

We finished the tactical training part of the syllabus in January and were preparing for the flight portion when the H-2 fleet was downed because of a problem with the door rollers. The rollers were not functioning properly, and the door would come off the aircraft. This had happened with disastrous results. The immediate solution was that you had to fly the aircraft with no doors, or the doors open all the way. That was not going to be possible in Norfolk during February, so the decision was made to send us to Key West, Florida, to do the flight portion of the syllabus.

The class really bonded during the Key West experience. It was us against the instructors. It did not help that some of the guys in the class could fly the aircraft better than the instructors. Also, we had a single vehicle assigned to us, so we all had to go everywhere together. We would have breakfast, pile into the van, and head to the air station for the day. They also made us responsible for running the hot refueling of the aircraft (refueling while the engines were running, and the rotors were turning) and crew swaps. This entailed ensuring the fuel truck, and the student was there when the aircraft pulled in for gas and a crew swap. The goal was to keep us as busy as possible to keep us out of trouble. They kept us busy, but we did enjoy ourselves as much as possible.

After being taught that it was impossible to have a complete electrical failure in the H-2, we were sitting on the flight line one evening when we heard this noise coming from the darkened taxiway. Lieutenant Dave Youngblood taxied out of the darkness with no lights on at all. Turns out you could have a complete electrical failure when the AC/DC generators and the battery decided to fail simultaneously.

Then there was the incident when one of our most experienced classmates, Tom Freeland, was shooting an autorotation, and he suffered a lateral trim strut failure. The trim strut was used to take out the lateral force of the cyclic. The cyclic immediately wanted to go all the way to the right. Tom was wrestling with the cyclic to keep it from going instantaneously all the way to the right stop. All this resulted in a "shaky" autorotation recovery, and the instructor pilot couldn't help but say, "that was a pretty poor auto for a guy with your fleet experience!" Tom asked the instructor to take the aircraft for a moment. As soon as Tom released the cyclic, the aircraft began to move drastically to the right. The instructor screamed, "holy shit!" and tried to center the cyclic. Tom just smiled at him.

It was eaten, sleep, and fly. On those few days when we did not fly, the senior guys would arrange for us to see the sights. We did a lot of "studying" at Sloppy Joe's Bar, especially when we heard it was Hemingway's favorite watering hole. We used to muse about being in Hemingway's town, at Hemingway's Bar, drinking Hemingway's drink. I had sixteen flights over twenty-two days in February to complete the flight portion of the FRS. I got to be a copilot in one of the aircraft flying back from Key West to Norfolk. Five H-2s in formation made quite a sight as we returned to the squadron. It was kind of a "mini cruise," and I really enjoyed seeing Linda and the kids again. We had to wait around for another month before we could go to the boat to complete our shipboard qualifications. I got to spend a part of that time in Survival, Escape, Resistance, and Evasion (SERE) school.

You learn a lot of things about yourself in SERE school, and not all of them are good.

SERE school was not as bad as people made it out to be. It was worse. I was even going to get a "two-fer," SERE School and Cold Weather Survival Training. The escape and evasion parts are difficult when you are wearing snowshoes. SERE school really does teach you a lot about yourself, mainly that you never want to get captured. I was the second senior man in the class, so I got to be the XO. I did well in the evasion portion of the syllabus. I made it to the first objective, and a cup of chicken soup never tasted so good.

While I was making my way to the final objective, the instructors told us to come out of the woods. We piled into a truck and were soon in the hands of the People's Republic of North America guards. I was at the bottom of the truck, and so as the others got out, I heard their initial interrogation. When I got out of the truck, they started asking me a lot of questions that I answered with my name and rank. They asked me what branch of service I was in, but I was not thinking very straight and decided that the longer they worked to get my branch of service, the less likely they were to interrogate me about other things. I soon found myself in March, in Maine, wearing nothing but my boots. The water seemed especially cold as they poured it over me. They must have been on a tight timeline because they quickly tired of this, and I found myself back in the truck. When we got to the camp, they asked me again what service I was in. "Do you think that we are so stupid? You are in the Navy. What other service has the rank of Lieutenant O-3?" My answer of the United States Coast Guard earned me the designation as an 'insolent one" and more special treatment. I also found out that they did not communicate with one another very well. Right after I got into the camp, the Senior Ranking Officer told me I should tell them my branch of service. I told the guards what my branch of service was, and they returned me to my cell. About ten minutes later, I was in my cell when "Igor" came in, and before I could say a word, I was up against the bulkhead, and he was screaming what branch of service are you in? I could see the look of disappointment in his eyes when I replied U.S. Navy. He left me in a heap on the floor of my cell.

It is hard to describe SERE school and what the lack of food and sleep does to a person. We all want to believe that we can hold out forever and not provide any information, be a John Wayne or Rambo. A few days at SERE school hardly compares to the years of torture and starvation that Navy Aviators and Naval Flight Officers went through in North Vietnam. At least for me, a few days into the experience, I began to wonder if I had not somehow been actually transported to the Peoples Republic of North America. You are in a mental fog and not thinking clearly. My worst moment came when I was acting as the CO. I was taken to an office, thrown on the deck, and told I must sign a book that documented my interrogation. I never thought about the possible consequences of signing the book. They took my signature and put it on a document where I confessed to a series of war crimes. The guards read the statement to the other prisoners. One of the prisoners said the statement was false and no Lieutenant in his Navy would be stupid enough to sign a statement like that. It must be a fake. The others also started yelling that it was a fake. They later told me about the incident, but it did not make me feel any better about myself and how I had failed to do the right thing when put in a situation to test my moral courage.

Detachment Alpha was currently deployed in the Mediterranean. We would do our relief of the current crew while in port at Athens, Greece. I was excited about the deployment and the opportunity to fly the aircraft in an operational environment. The detachment aircraft was an HH-2D that had a different rotor head system and slightly different operating limitations, no anti-submarine gear, and sliding cargo doors on both sides of the aircraft. The principal difference between the "Foxtrot" model that I had flown in the FRS and the "Delta" that I would fly while deployed was that the tail wheel was further aft on the HH-2D than it was on the SH-2F that I had previously been trained to land aboard ship. That meant you had to be a little more conscious of where you put the tail wheel on landing. The deck on the USNS HARKNESS (TAGS-32) was big, as was the aircraft hangar. This ship design made it easy to move the aircraft in and out of the hangar while also providing a lot of space to stage the external loads that we would be hauling to the support sites ashore. We also could winch the aircraft into the hangar as opposed to having to move it by hand, like I would learn to do on Frigates and Destroyers.

The Navy Oceanographic Unit (OCUNIT) that we were assigned to support was tasked to map the depth and contours of the sea floor on the northern and southern approaches to the Suez Canal. To do this, the ship had to be very precise in its navigation. The detachment was responsible for logistically supporting the

OCUNIT's establishment of navigation aids at remote sites ashore in Egypt. Radio beacons were established ashore at known geographic positions. The ship would then triangulate between the sites to know where it was exactly on the surface of the earth while mapping the ocean floor below. These sites were manned by personnel who ensured the equipment was operating properly. Our job was to lift the personnel, equipment, and supplies necessary to set up and support these sites.

There were several interesting variables in servicing these sites. First, there was the sand. This was before the Gulf Wars, and experience flying in that environment, especially by Navy pilots, was limited. Fortunately, we learned quickly how to manage the brownout conditions that we encountered on most landings at the sites in Egypt. You did a precision approach to a no-hover landing. Once you were in the brownout, you kept the rate of descent constant until you impacted the ground. Flying an aircraft that was built to absorb extreme shipboard landings helped.

The other thing you needed to consider was the minefields. While our landing spots had been cleared, there were hundreds of thousands of mines in the desert left from various Egyptian-Israeli conflicts that had never been mapped, recorded, or removed. That fact was driven home by some of the spots we operated from where you could actually see the mines where the dirt had been blown away. If you had an emergency of any sort, you had to fly the aircraft until you could get to one of these cleared spots.

When we deployed, the Egyptians and the Israelis had not yet signed the Camp David Accords, and the border between them was the Suez Canal. We had two passports, one to use for Egypt and one to use for Israel because they would not accept one another's. Evidence of the 1973 Yom Kippur War was evident along both sides of the canal in destroyed anti-aircraft sites and armored vehicles. As we did most of our flying in Egypt, we had an opportunity to observe first-hand what the Israeli Air Force and Army had done to existing Egyptian positions. Port Suez in the south had suffered extensive damage during the war. We were always aware that we were between two enemies who could start combat operations at a moment's notice. The pirate radio station in the Mediterranean, "The Voice of Peace," regularly played a station break that featured Anwar Sadat and Menachem Begin. Sadat says, "no more war, no more bloodshed," and Begin is heard to say, "this should be the last war." We sure hoped that was true.

The United States Naval Ship (USNS) designation meant that the ship was staffed by civilian mariners from the Military Sealift Command (MSC). My first indication that life was different aboard a USNS ship was wearing tennis shoes when I was not flying. There was also a ship steward who took care of our rooms and laundry. We were required to pay for our meals, but unlike on a Navy ship, you only paid for what you ate, and you had two to three entrée choices at every meal. We also had to get used to the idea of normal working hours and trying to avoid overtime for the civilian crew. The MSC folks were thoroughly professional and knew their stuff. The Master was a crusty sort who had been torpedoed during World War II and was the guy who sailed the Mayaguez out of Cambodian waters after it had been retaken. For some reason, he loved naval aviators. He told the OINC if we ever had a particularly difficult day and we needed a little "liquid courage," he would be happy to provide us with some. Our mission only required us to fly in the day and visual meteorological conditions, so the likelihood of us needing a drink was remote. But our schedule did allow us to regularly attend the movie each evening in the wardroom with the Master.

The detachment consisted of three pilots. The Officer in Charge (OINC) Lieutenant Dave Youngblood, Detachment Maintenance Officer (MO) Lieutenant John Burton, and me, the frigging new guy (FNG). I was the Division Officer, Administrative Officer, and Operations Officer. AD1 Windham was the crew leader, and the rest of the detachment were AE2 Waak, AEAN Fontenot, AD2 Cook, AT3 Temple, and AMS2 McQueen. Petty Officer Cook and Airman Fontenot were our aircrewmen. I did the daily scheduling of the aircraft in conjunction with the Oceanographic Unit, managed all the message traffic, and made sure our folks were up to speed on advancement and training. The OINC had a stateroom of his own. I roomed with John Burton, and being the junior aviator, I slept in the top bunk.

We had gone through a couple of periods on the line when we learned the ship would be going to Sicily for liberty. This would give us a chance to fly into Naval Air Station Sigonella to get needed aircraft supplies, calibrate some of our equipment and pick up an engine. I remember getting up before sunrise on the day we pulled in and being able to smell the land prior to the sun coming up. My job was to load a truck on the pier with all the gear that needed to be calibrated, double-check our required supplies, and get the det members who were going to assist me into the truck. It was great being ashore again, and all I really wanted was a cheeseburger and a chocolate shake. I was also looking forward to being able to read the International Herald Tribune and the Stars and Stripes.

Time flew as we got the gear inducted, and I went to the supply warehouse with our list of required supplies. I found everything we needed. The truck was loaded, and we took everything back to the ship and secured it in our spaces. The next day we were scheduled to transport the aircraft engine from the Aircraft Intermediate Maintenance Department to the HARKNESS as an external load on a sling attached to the aircraft's cargo hook. The engine was supposed to be ready in the early afternoon, and we expected that we would be landing on the ship before sunset. Things did not work out that way.

The folks in the Powerplants Shop at AIMD Sigonella were taking longer than anticipated to get the engine ready for us to transport. It was becoming obvious that we were going to run into a crunch with nightfall. We were supposed to be flying daytime only, and we certainly were not certified for a night vertical replenishment (VERTREP). The engine was ready about sunset, and the OINC asked the aircrewmen and me if we had any qualms about doing the vertrep mission at dusk. We all said we were good to go.

They finally had the engine staged for us to pick up when it was well past dusk. The engine was staged on one of the helo pads adjacent to the runway. We pulled into a hover over the engine, and the ground crew hooked us up. The aircrewmen said everything looked good. Dave Youngblood pulled power to put tension on the pendant that attached the aircraft's external cargo hook with the engine canister. As he continued to increase power, I could feel the weight of the engine canister coming onto the aircraft. I called all gages in the green, and the crewman cleared us for forward flight. We had traveled about fifteen to twenty feet when my side of the aircraft lit up like it was daytime, and I heard this loud "boom." The left engine had suffered a massive compressor stall, and we lost all power on that engine. Dave Youngblood and the two crewmen simultaneously called for me to "pickle" or release the external load. I hit the button and simultaneously reached for and actuated the manual external release. When Dave landed the aircraft, he asked me to get on the rudder pedals with him. "Do you feel that vibration," he asked. I said yes, and we went through our shutdown procedures. When we got out of the aircraft, we saw that the external load had not been released. The tail of the aircraft was sitting on top of the engine canister with the pendant still attached to the aircraft. If we had gone any further, we would have dragged the engine canister along the ground, and we would have pitched forward and struck the ground. We never could figure out why the electrical button or manual lever did not function properly and release the load. We may have been too low to the ground for the hook to open completely. The "gouge" on doing external loads in the community was that you never armed the cargo hook release and should always release the hook manually.

Thus, began our unscheduled two weeks in Sigonella. There was a rush to get the detachment, their personal effects, and all the equipment that we would need to make whatever repairs were needed to the aircraft from the ship to Sigonella. The ship had to leave that night and return to Egypt for their scheduled line period. They estimated a return in two to three weeks to pick us up. I was busy trying to get accommodations for the detachment. Dave Youngblood was busy getting hangar and working spaces. We would do our maintenance in the hangar, and we had a trailer to use for the detachment's administrative spaces. There were no rooms available on base, so we wound up staying at a hotel that the base contracted for newly reporting personnel to use when they were transferring to Sigonella on Permanent Change of Station orders.

The damage to the aircraft required an engine change and the replacement of the entire tail pylon. Fortunately, AIMD had another engine, and we could not use the one we had landed on or the one that blew up. Supply just happened to have a tail pylon for an HH-2D. We settled into a routine to methodically make the aircraft whole again. We worked twelve-hour days, would find someplace out in town to eat, and then return to the hotel. Our crew leader AD-1 Windham had about 18 years in the Navy and was very experienced. About half of the rest of the det was on their first cruise. We all took notice of how our seniors went about their business and adjusted accordingly. It took us about a week to get all the required maintenance done, and then the OINC said we would fly one hop a day for pilot proficiency and practice the emergency procedures that we could not do while on the ship.

HSL-30 DET ALPHA

FIRST ROW L-R AEAN FONTENOT, AD2 COOK, AT3 TEMPLE, AMS2 MCQUEEN

SECOND ROW L-R AE2 WAAK, LT COYNE, LT YOUNGBLOOD, LT BURTON, AD1 WINDHAM

As the third pilot on the detachment, I would normally expect to fly alternately with the two aircraft commanders. While we were in Sicily, Dave Youngblood would fly with me, and then we would swap pilots, and he would fly with John Burton. Aboard the ship, the pilot who was not flying would man the control tower and coordinate all the movements of the flight deck crew and the ship's company. When I had the duty, I could notice that there was a difference in how the other two pilots flew. Dave Youngblood was very calm, which was evident in the smooth control of the aircraft and his tone on the radio. When John flew, the movements were not as well controlled, and he sounded more excited on the radio. While in Sigonella, our maintenance trailer was close to the helo pad. One day when Dave and John were flying, I was inside the maintenance trailer when I heard the detachment members outside the trailer laughing. They were guessing which pilot was flying the aircraft. In one case, the aircraft was rock steady over the pad. Then there would be a momentary pause, and then the aircraft would begin to drift a little to the right. "Looks like the MO is flying!" they said. After a few minutes of watching, I would hear, "the OINC has the stick now!" The last few days in Sig, I did not fly at all as Dave and John spent a lot of time practicing basic air work in the pattern at Sigonella.

My next two emergencies in the aircraft would involve a hydraulic pump line and that annoying cargo hook again.

We were doing a logistic run to one of our remote sites at Damietta, west of Port Said, Egypt, and the ship was about thirty miles off the coast. We noticed a jump in our hydraulic pressure, and then the pressure began to fall toward zero. One of the crewmen looked out and said he could see fluid on the side of the aircraft. Our aircraft was painted white to make it appear less threatening, and the red hydraulic fluid stood out as it ran alongside and down the tail of the aircraft. The H-2 was unique in that the flight control system was mechanical with hydraulic assistance. There were a series of cranks and rods that connected the flight controls to flaps that were located on the main rotor blades. When you moved the cyclic, the control input was manually transmitted to the flaps that would change the angle of attack on the blades, and you would turn, climb, or descend. While you could fly the aircraft without the hydraulics, it was like wrestling with a five-ton bull. There was also the issue that the hydraulic pump was lubricated by the hydraulic fluid, and with none in the tank, the pump might overheat and cause a fire. We told the ship we would be landing at the remote site, and after we had a chance to inspect the aircraft, we would contact them with our proposed plan of action.

The landing was uneventful, and when we shut down, the crewman discovered a hydraulic line had ruptured. We notified the ship of our problem. The crew leader got back to us and said they would fabricate the line and send it ashore by small boat along with a mechanic to fix it and replacement hydraulic fluid. There was nothing else we could do that day, and the guys at the remote support site recommended we go into Port Said to spend the night. We found a hotel and then went to the United Nations Club. The Club was in the basement of a house that the UN was renting that they had made into a bar. As I recall, drinks were about ten cents apiece. We had been at sea for a while, and sailors being sailors, we did our best to consume copious quantities of alcohol. We did not make the 12 hours bottle to throttle rule that most naval aviators are required to observe. The ride back to the remote site from Port Said was especially hot and bumpy. For the first time in my life, I really wanted Dave Youngblood to do the flying so I could catch a nap. Of course, Dave told me that I would be flying and expected to make the landing aboard the ship.

Later in the cruise, we were tasked with transporting a one-thousand-pound generator from Damietta to the ship. Everything about the flight was normal until it was not. We moved over the load and were hooked up. The weight came on the aircraft, and the gages were all in the green. Having been fooled once in Sigonella, we did not arm the cargo hook. We transitioned to forward flight and were climbing out at about 750 feet and about two miles off the coast when the aircraft suddenly got lighter. You could feel it in the seat of your pants. I knew that the generator was headed for the bottom of the Mediterranean. Dave Youngblood rolled the aircraft into a turn, and I could see the magnificent splash that the generator had made. Fortunately, no one was hurt, and we returned to the ship and filled out the incident report.

I got better and better at flying the aircraft and became comfortable with hauling external loads. The most challenging was the pallets of color television sets we would lift into the Port Said Duty-Free Zone. We had Egyptian Liaison Officers on the ship to coordinate everything that we did ashore and along the coast. They found out they could order color television sets through the ship's store, and then we would transport them ashore when they arrived. They would not have to pay duty on the TV sets because of the Duty-Free Zone. You must be very gentle when flying with a pallet loaded with tv sets, especially when you set them on the deck. It's funny, but we never had any problems with the cargo hook when we were hauling the television sets.

About four months into the cruise, the OINC told me that the MO would be rotating home, and I would become the new MO. I was surprised as the pilot coming out, Lieutenant Dave Palleson, was senior to me and would become the OINC when we rotated crews. I turned over my OPS/ADMIN duties to him when he arrived and settled into the MO job.

Dave Youngblood was very patient with me and some of my more impulsive traits. I spent the entire cruise trying to convince him we could do a touch-and-go on the road between Port Said and Damietta, and no one would ever know. He would shake his head no and just smile. He patiently corrected my mistakes on forms and reports while encouraging me to learn as much as I could about the aircraft. We discussed my accident in the training command in detail, and he still could not believe that the alarm bells were not going off earlier for me during the flight. I have never been a big believer in mentors, but he was one of those guys that I learned a lot from. He was very calm on the ground and in the air. He was thoroughly professional in everything he did. I would draw upon the lessons I had learned with him when I got my own detachments.

I do not know if it was a coincidence or not, but on three of my cruises, I had Aviation Machinist Mates (ADs) as my crew leader. I always figured it was because I was a slow learner, and they could explain things slowly and simply so that even I could understand them. I had noticed that there had been friction between our crew leader AD-1 Windham and the old MO (LT John Burton). John provided a lot of detailed guidance to Petty Officer Windham, and I thought that he was overdoing it based on my previous experiences working with senior enlisted personnel when I was stationed in Bermuda. Not long after I became the MO, we had an inflight incident, and we had to make out a safety report. I went to AD-1 Windham and asked him to fill out all the technical information for the report, and I would do the narrative section. He looked at me and said the old MO had done that. I said, "you have 18 years in the Navy, and you are telling me that you don't know how to fill out this form?" He smiled, took the form from me, and said, "you have been watching." We got along great after that. We worked well together as a team, and I learned the complexity of the different maintenance evolutions at the Organizational Level and, more importantly, how long they took and what could go wrong. He taught me a lot about the H-2, the Naval Aviation Maintenance Program (NAMP Instruction 4790.2), and how to run a detachment.

My time as the MO flew by, and before I knew it, I was scheduled to return to the squadron. I would be returning individually, so I arranged to take leave en route and meet Linda in London. I was delayed getting to London on my flight from Athens, Greece, via Paris and wound up spending a night in Paris worrying about how I was going to connect with her when I arrived. There were no cell phones then, and our plan was originally to meet at the airport. I did not know what hotel we were booked into, and the fallback plan was for her to notify the naval attaché at the embassy about where she was staying, and I would check in with him when I arrived. It took me all day to get from Paris to London, and Linda was at the airport wondering if she should just go back to the states. I got off my flight and was heading to baggage claim while running through my mind what I would do next. I heard someone calling my name, and there she was. We made our way to the hotel and tried to catch up on what had happened at home with our daughter over the previous six months. After a couple of days, I surprised myself and asked her to cut the week short because I wanted to get home and see our daughter Carey. I needed to go home.

When I returned to the squadron, I was assigned to the Administrative Department as the First Lieutenant and Public Affairs Officer. The First Lieutenant oversees the maintenance and upkeep of the squadron facilities. So, I was the plumber, electrician, and cleaning hangar guy. There was not a whole lot to the job except for ensuring that the microphone worked when the squadron held quarters. The Public Address system was a constant source of trouble. No matter how many times I would check it out, the system would always fail when the CO was about to use it. We never quite figured out why. It might have been the electricians in the squadron having a little fun at the First Lieutenant's expense.

The Public Affairs Officer (PAO) responsibilities were the easiest job I ever had in the Navy. Having started in Radio/Television Broadcasting and Journalism, it was easy to turn out a couple of articles a month for the Base newspaper. I also made sure that all our folks who received awards had a notice sent to the Fleet Hometown News Center. I was able to do my ground job and still try to learn as much as I could about the H-2.

The Det Alpha pilots normally taught the utility portion of the FRS syllabus, or as we used to call it, "Rocks and Blocks." We trained the pilots on how to do external loads by picking up huge concrete blocks and then flying around the traffic pattern at the Naval Air Station Norfolk heliport. Most pilots were initially hesitant about hauling the external load but would get better over the course of the hop. I made sure to get the load swinging so that they could learn how to steady it by decelerating and descending. You can tell what the load is doing by the feeling in your butt. You also want to make sure that the load does not get out of balance aerodynamically. I once had to do an external load in Egypt at about thirty knots for thirty miles because if you went any faster, the external load would go aerodynamic.

I asked to go through the FRS instructor syllabus so that I could also fly familiarization and instrument flights with the students. Because I didn't have a previous tour in H-2s and therefore didn't have a lot of H-2 flight time, there was some reluctance on the part of the operations folks to let me do these other types of hops. I said I had been an aircraft commander in the UH-1N and had as much total flight time as any of my contemporaries that were coming from the fleet squadrons for an instructor tour. Operations agreed that if I made it through the instructor syllabus and was recommended by the standardization check pilot, then I would become a regular instructor. While I was going through the instructor syllabus, I was also doubling as the duty functional check flight copilot. As such, I got to fly with the pilots assigned to the maintenance department, the folks that knew the most about the aircraft.

As the duty maintenance functional check flight (FCF) copilot, I learned a tremendous amount about the aircraft and actually flying the functional check flight. I learned that not everyone is cut out to do FCFs. Good FCF pilots take the aircraft right to its limitations to ensure that the maintenance has been done properly and the aircraft is safe for flight. This means that you are going to be uncomfortable doing some maneuvers. I learned a lot from a former Det Alpha pilot Tony Montemarano. The first time I was his copilot, I noticed him go from the aircraft that we were doing our FCF on to another aircraft beside it on the flight line. He got up and checked something on the rotor head of the second aircraft and then called for a troubleshooter. Tony got up on our aircraft with the troubleshooter and asked, "is this component supposed to move?" The troubleshooter said, "no." Tony said, "well then, you better fix that other aircraft because it does move on it." I learned the aircraft comparison maintenance technique. The squadron troubleshooters hated it whenever Tony asked a question because they knew there was going to be a 50-50 chance that they would end up doing additional unscheduled maintenance after Tony had completed his aircraft comparisons.

Once I made it through the instructor syllabus, I began to get hops other than "rocks and blocks." But a pattern soon developed that I was being assigned to do what we referred to as the "doppler hops." These consisted of doing simulated search and rescue pickups with the H-2 in day and night conditions. Day dopplers were a piece of cake, but night dopplers were a different matter. While the doppler system in the H-2 was much better than anything I had flown in the H-1, most pilots did not fully trust it, especially at night. So as the junior instructor pilot in the squadron, I began picking up more of the night doppler hops. I would usually have two students, and I would day qualify the first and return to the squadron to swap out copilots and then return to get the second pilot day qualified. The trick was to get the second pilot day qualified so that you could immediately begin the night dopplers. To qualify as a night doppler, the maneuver had to occur after official sunset. As everyone knows, it does not get dark immediately after the sun sets, and naval aviators refer to this as "pinkie time." Pinkie time is the time between when the sun sets and it begins to get really, really dark, and it counts as nighttime.

I would get the second pilot qualified when it was still "pinkie time" and return to the squadron for the first copilot. The senior pilots who were requalifying would do their day dopplers as the sun was setting and then their night "pinkies." I would return the senior pilot to the squadron and get the junior first pilot, who had never done a night doppler in the H-2, and we would then return to finish the night dopplers as it got dark. The senior pilots going through the FRS for requalification loved doing the "pinkies," and the frigging new guys did the "dark darks." My one iron rule at night was to never back down. This would occur if you overshot the flare that we put in the water as a reference to simulate where the survivor was. If you went by the flare, I briefed that we would keep on going and come around for another approach. At night when you try to back up, there is a tendency to also descend. This can have seriously dangerous consequences when you are only 25-30 feet over the water.

A pattern developed where I would do a lot of maintenance check flights as a copilot until the FRS class got to "rocks and blocks" and dopplers. That sequence culminated when I did an entire FRS class of eight students for their day/night dopplers. They awarded me a plastic map of the area in which we had done our dopplers and the honorary nickname of "Pinkie" because I had worked so hard to get as many qualified as possible during "pinkie time." My motto at that time was I was the only one that was going to scare me at night because I would not let the students get that far.

Naval Aviators are famous for their call signs, but I never had one that stuck through my various squadron tours. In HSL-30, I was known as "Pinkie Coyone." This nickname represented a gradual change over time. One day on the flight schedule, they spelled my name wrong as "Coyone," so for a while, that was what I was called. Then I became "Wiley Coyone" because I used to howl like a coyote during the required radio checks while doing the night dopplers. After the FRS class dubbed me "Pinkie," I finished off my time in the squadron as "Pinkie Coyone."

I got a phone call at home one weekend, and the Squadron Duty Officer (SDO) said I was to report to the squadron immediately. I asked if I needed to bring my flight gear and a change of clothes. He said yes. I thought something had gone wrong with the detachment, which was still deployed. A more seasoned aviator would have deduced the obvious; there had been an accident.

When I arrived at the squadron, I noted there were a lot of cars in the parking lot. When I went in, I was immediately asked why I was there. I said the SDO had told me to report to the squadron. I was told to go up to the operations office. I learned that one of our aircraft on a cross-country flight had crashed, and there were no survivors. I was going to be a member of the Accident Mishap Board (AMB).

The AMB had four members on it. The Operations Officer, Lieutenant Commander Diener Scango; the Assistant Safety Officer, Lieutenant Sonny Pettit, the Maintenance Quality Assurance Officer Lieutenant Tony Montemarano; and me, as the Administrative Officer. Tony and I were both current in the HH-2D, the aircraft that had crashed in New Hampshire when it was on a cross-country flight from Naval Air Station Brunswick in Maine back to Norfolk, Virginia.

We got to the accident site and began to examine the wreckage and formulate our Plan of Action (POA) for how we were to conduct the investigation. After about an hour, the Ops Officer called me over. He said I am going to ask you to do the toughest job you will ever have in your naval career. I want you to do the identification of the bodies. The Ops officer did not know it, but I had some previous experience with dead bodies. On one Christmas break from college, I had worked part-time as an ambulance driver for my older brother's father-in-law. I had several instances where we would go into a home to pick up a deceased member and transport them to the morgue, where I had the opportunity to see a lot of dead bodies in various stages as they went through autopsies. So, the sight of a dead body did not bother me; the fact that I knew them did. The aircraft had crashed and rolled down a hillside before it caught fire. One individual had been thrown clear and was easily identified. The rest had been severely burned in the fire after the crash, and it was much more difficult to identify them. It was traumatic for me at the time as I had worked with one of the pilots and played tennis with him. He introduced me to the K-SWISS tennis shoes that I wear to this day. Every time I put them on, I think about him.

We pieced together as much as we could about the flight. There were three experienced H-2 pilots, an aircrewman, and a plane captain. The flight from Brunswick to Norfolk would be mostly in Instrument Meteorological Conditions (IMC), but the weather forecast said the clouds would bottom out at one thousand feet, and they would be able to fly Visual Flight Rules (VFR) beneath it. However, the forecast was wrong. The clouds stretched from their assigned altitude right down to the ground.

They were proceeding on their route normally when they began an immediate drastic descent. There was no radio call declaring an emergency nor a request to leave their assigned altitude for a lower one. Based upon the time between the loss of radar contact and the time of the crash, they flew around at tree top level for several minutes before the crash. People in the neighborhood described the noise of a helicopter directly over them, which caused things in their houses to start shaking. They also said that they could not see anything because of the fog. We speculated that while flying around looking for a place to land that they had sighted a potential landing site, and in maneuvering the aircraft around, they lost sight of the landing spot in the fog and flew the aircraft into the ground.

The AMB tried to find an emergency that would be drastic enough for them to do an autorotation in instrument conditions but did not allow them to get out a radio call. The squadron sent another HH-2D up to New Hampshire, and Tony and I tried to simulate the flight profile the accident pilots had flown in their final seconds. The only way we could simulate their rate of descent was in a full autorotation with the aircraft out of balanced flight. We could not find anything wrong with the aircraft mechanically. Engines and rotor systems performed normally. There might have been a problem with their radios on a previous portion of the flight, but they were in contact with the center controlling their flight prior to the crash with no problems. We could not say that there had not been an electrical problem or a momentary chip light, but we could not prove it either.

The squadron had a memorial service while we were still up in New England conducting our investigation. Once we had completed all our fieldwork, it was time to return to the squadron. Tony Montemarano and I would fly the aircraft back. I did not realize it at the time, but there was a lot of unease in the squadron about flying the HH-2D because we had not been able to find a reason the aircraft had crashed. I did not feel that way and had no problem with it, but the decision was made that we could only bring the aircraft back under Visual Meteorological Conditions (VMC) or Visual Flight Rules (VFR). Our trip home was being delayed by the same type of weather that had been present when the crash occurred. We were able to get out special VFR but soon ran into a low ceiling and found ourselves hunting for some clear spots to get back to the airport. Once we got back, we both confessed that we were doing the same sorts of things that got the previous aircrew in trouble. We decided to stop for the day. About an hour later, there was blue sky and sunshine, but we had already decided to stop for the day. Our subsequent flight back to Norfolk the next day was uneventful.

Not long after the AMB concluded, we got word that Detachment Alpha would not be deploying again for a while because the USNS HARKNESS was going into the shipyard for an overhaul. The squadron was talking about disbanding the detachment until the ship would be ready again. I could have cut my sea duty short and stayed in the squadron on shore duty as an instructor pilot, but I still had that desire to get to the real Navy and operational LAMPS Squadron in "The Fleet." I called the detailer and asked whether I could go back-to-back sea tours vice staying in the FRS on shore duty. He asked, which squadron do you want to go to? I did not want to move from Norfolk. I did not know anyone in HSL-32, and three of my fellow RAG classmates were in HSL-34. I would become a Greenchecker.

CHAPTER VI

THE FLEET!

HELANTISUBRON LIGHT THREE FOUR (HSL-34)

THE GREENCHECKERS

JANUARY 1980 – JUNE 1982

There was an entirely different feeling to checking into HSL-34. Coming from the FRS, everyone assumed I knew a lot about the H-2. I went through all the required flight checks to become a Helicopter Aircraft Commander, Functional Check Pilot, and Assistant NATOPS Officer. I also became the detachment maintenance officer for Detachment One.

I had gone through the FRS with my new OINC Ed Driscoll. He started as a surface warfare officer on the USS INDEPENDENCE, then became a naval aviator and did his first aviation tour at Naval Air Station Midway. He was very intelligent and dynamic. I was drawn to his positive philosophy about how to work in the Navy. The detachment nicknamed him "Tasmanian Ed" after the cartoon character that was constantly spinning. He was all energy and curiosity and did not take no for an answer. We set the goals and objectives for the detachment and made the det a part of that process. Within a couple of months, we had the most flight time and the best operationally ready rate in the squadron. We would schedule cross-country flights on the weekends so the det could launch us early on Friday and have the rest of the weekend off. We established our reliability in achieving any mission that we were assigned and found ourselves becoming the go-to detachment for important operational commitments as well as the static display aircraft at important Navy events. We went to the Gulf of Maine onboard the USS PETERSON to conduct anti-submarine warfare exercises with a U.S. nuclear submarine and more than held our own during the operation.

I was sad to learn that we would not deploy on a long cruise with Ed. Our new OINC Lieutenant Commander (LCDR) Bill Gibson, or "BG," would be returning from a cruise as a detachment maintenance officer and taking us to sea as our new OINC. He would be going on back-to-back long cruises. BG was very calm and measured in the way he ran the detachment. He had been a detailer and was used to considering a larger view of circumstances and situations. BG had been selected early for Lieutenant Commander and was considered on the fast track to bigger and better things. He balanced out my tendency to react without thinking and go off half-cocked. He taught me a great deal about the politics of the Navy and how the detailing and placement systems worked. He was the kind of guy you hated to let down because he trusted you to get the job done and do the right thing. Lieutenant Rob Scott was the third pilot and took care of operations and administration.

The detachment was comprised of AMHC Dusty Hoffman as the crew leader, and the rest of the det were ATAN Ellis, AN Council, AE1 Myers, AD2 Van Putten, AEAN Fralix, AMH2 Broten, AW2 Jennings, AD3 Ratliff, AX1 Perry, AMHS Wethington, and AWAN Hood. Petty Officer Jennings and Airman Hood were our aircrewmen.

We would be going to the Persian Gulf onboard the USS PETERSON DD-969, a Spruance Class Destroyer. We had worked with them before in the Gulf of Maine. The "Plank Owners" of the ship's crew were turning over with the CO/XO, and the major department heads who had commissioned the ship were leaving for new assignments. It was the first time that I experienced how a command could change so drastically under new leadership. The earlier CO of PETERSON saw the detachment as an asset that made his ship more combat capable. I learned that this was not the prevailing view among many officers in the surface warfare community. Most ship COs felt that one of two things could happen with an embarked LAMPS detachment, and one of them was not good. So, they tended to try to micromanage or limit the detachment's operations. The new CO of the ship had an engineering background and demanded that the OINC explain any maintenance actions using the H-2 Maintenance Instruction Manual or MIMS. Before BG would talk with the ship's CO, Chief Hoffman would open the MIMS and go over everything with the OINC, who would then brief the CO in detail.

The ship department heads took their cues from the CO and made everything we tried to accomplish more difficult. When the LAMPS detachment is not embarked, their aviation spaces (hangar, office, and supply spaces) are occupied by the ship's company. When the det moves aboard, the process of evicting the black shoes (surface warfare folks) begins. We had been aboard for about two weeks, and our maintenance workspace had a torpedo dolly that was used to move torpedoes within the ship's magazine. It should have been moved before we embarked, and the Weapons Officer said he would do it but never got around to it. I had three or four conversations with the Weapons Officer, and nothing was done. My frustration boiled over, and I confronted him about the dolly. I was shouting at the top of my voice and had gotten to the point where I was talking about his mother and her sexual habits when he held up his hand and said, "I'll move the dolly. I didn't know that you were that serious about it." Surface Warfare Officers have a tremendous amount of responsibility and a million things they are held accountable for. As a result, they tended to regard our requests as background noise unless you were emphatic about getting it done sooner rather than later.

The ship departed Norfolk for the Persian Gulf, steaming independently. That is, we were not going with a carrier or a group of ships. We did rendezvous with the USS JESSE L. BROWN (FF-1089) out of Charleston and did a series of exercises between the ships. The JLB did not have a LAMPS detachment embarked, so we worked their flight deck crew up with deck landing qualifications and refueling our aircraft. One day we were flying and received a radio call from the JLB that they had a man overboard. We were doing an anti-ship surveillance and targeting mission, so we were not rigged for search and rescue (SAR). We did not have a second crewman or any SAR gear onboard. I alerted PETERSON and requested they set flight quarters and notify the det that we would land, load our wet swimmer and his equipment onboard and then conduct the SAR.

We proceeded to the ships reported position at max airspeed and got as much information as we could about what had happened. We sighted the survivor, and I decided to try and do the pickup without the wet swimmer. I did not want to lose visual contact with the survivor and risk something happening to him while we landed on PETERSON. My aircrewman talked me into a hover over the survivor and lowered the rescue hook and sling or horse collar into the water. He told me that the survivor was swimming to the horse collar and that he had gotten into it but was hooked up backward. "LT, what do you want me to do?" the aircrewman said. I asked, "is he in the horse collar?" The crewman reconfirmed that he was, but that instead of having the horse collar supporting his back, he had it under his chest; he was in it backward. I told him, "Keep him coming into the aircraft!" I lowered my hover so that if the survivor came out of the horse collar, he would not fall that far. The crewman reported the survivor at the door and then went off mike as he wrestled the survivor into the aircraft. He came up on the ICS and confirmed the survivor was onboard and in decent shape. I let both the ships know that we had him onboard and asked the JLB whether they wanted us to return the seaman to them or take him back to the PETERSON. The CO of the JLB came on the radio, thanked us, and said they were setting flight quarters and wanted him back. After we landed and disembarked, the seaman, no worse for wear, the flight deck crew brought out a couple of gallons of ice cream for us. During World War II, when a destroyer would pick up a downed pilot and return him to the carrier, the carrier gave them ice cream as a reward for a job well-done, as well as an incentive to do so in the future. It was JLB's way of thanking us.

AT SEA RESCUE OF USS JESSE L. BROWN SAILOR

When we arrived back at the PETERSON and had shut down, the CO came on the ships 1MC (public address system) and thanked the bridge team for keeping track of the survivor, the ops folks for producing a rescue plan, the bosun's mates for readying the motor whaleboat for a rescue, the OOD for his expert conning of the ship and everyone who was ship's company for their outstanding performance. There was a slight pause, and he said, and the det. That about summed up our relationship with the PETERSON.

Cruising through the Mediterranean, the CO asked if we could fly the Ops Officer (OPS) and the Chief Engineer (CHENG) into Souda Bay in Greece while the ship pulled in for fuel at the NATO Piers. By flying them in, they would be able to call the states from Souda Bay and have some time to get updates on some of the things that the ship was trying to get done. We said sure, checked the publications, and sent a Prior Permission Request or PPR to the embassy in Athens asking them to liaison with the Greeks and get us permission to land at Souda Bay. We sent the message on a Friday, and the ship would be pulling in on Monday, so our antenna should have been up when we did not get a response. But we did not give it a second thought.

We launched and contacted the Air Traffic Control folks at Souda Bay about clearance to land. There was a long pause on the radio, and they asked us for a PPR number. We said we did not have it in the aircraft and that if they contacted U.S. Navy operations, they would have it. The tower came back and said the U.S. Navy did not have the PPR. We continued talking with the tower while our fuel got lower and lower. Finally, we said, "if you do not give us permission to land soon, we will run out of gas. The tower wanted to know if we were declaring an emergency for low fuel. We said not at this time, but it will not be long before we will declare an emergency. They cleared us to land. When we were shutting down, a vehicle rolled up, and a Greek Air Force Colonel got out and asked us why we had landed without permission. We told him about our PPR request. He said we had declared an emergency to get around the PPR. We said we had not declared an emergency but that it would not have been long until we would have been forced to do so. He thought about that for a moment and then said, "your aircraft is impounded. Do not attempt to leave until I give you permission." We went into base ops and asked what was up. The Operations Duty Officer (ODO) said the Greeks were extremely sensitive about the U. S. Navy using Souda without their concurrence and that the last time this had happened, the Greeks impounded an H-53 helicopter. I asked, "how long before they let him leave?" The ODO said it had been months, and then they had to disassemble the aircraft and ship it out on a truck. The pucker factor went up tremendously.

HSL-37 DET ONE ONBOARD USS PETERSON DD-969

THE AUTHOR IS IN THE PILOT'S SEAT

While we were trying to get someone in the U.S. embassy to come up on the net, OPS and the CHENG were getting a lot of stuff done. A couple of hours went by, and we still were not able to get anyone to approve our PPR request. I guess they did not have a Prior

Permission Forgiveness form to fill out. The ship had been scheduled for a Brief Stop for Fuel (BSF), and we were beginning to worry that they would refuel and be beyond our range before we had sorted this whole thing out. The Greek Colonel suddenly appeared and said we had five minutes to launch. We all bolted for the aircraft and turned up in record time. After we launched, we brought up the Tactical Air Navigation System (TACAN) to get an idea of how far away the ship was. The needle would not lock on, and we called the ship on the radio to get their position. There was no answer, and we thought, holy crap, they are out of radio and TACAN range. After about our third radio call, the ship's air controller came on and said they were pier side and had just started refueling. Their estimated time of departure would be in an hour and a half.

We told him we were airborne and would require a green deck for landing in one hour and forty-five minutes. About half an hour later, my bladder overpressure light went on. I thought about going back to Souda but quickly ruled that out. We could not go to the ship. Here is where my Det Alpha experience came into play. I called for the landing checklist and said I was going to land in the field ahead of us to conduct an external inspection of the aircraft. The H-2 does have a urination tube under the pilot's seats, but I had never quite mastered the ability to coordinate all the necessary actions to use it. Unzipping of the flight suite. Removal of intervening layers of clothes. Inability to position the tube properly for my undersized unit so that I would not spray the cockpit. I was willing to go to a Greek jail, but I was not going to urinate in my flight suit. We landed, and I performed the visual inspection of the aircraft and reset my personal bladder overpressure light.

Airborne once again, we checked in with the ship every fifteen minutes, but it did not seem they were ever going to complete the refueling process. Our thirty-minute fuel light illuminated, and the ship still was not done refueling. The light did not guarantee that you had thirty minutes of fuel. The ship finally said they were preparing to get underway. I told them to set flight quarters, and I would accept any winds they could get, but I would be landing as soon as they exited the fueling piers. I landed on PETERSON as they were in the channel departing the NATO fueling pier.

Our arrival in the Persian Gulf was routine. The ship/detachment fell into a daily routine. We conducted dawn and dusk patrols and were surveilled by the Iranians on a regular basis. One of their P-3s would pay us a visit just about every day, and their surface vessels would also come out and challenge us on a regular basis. We spent most of our time in the South of the Gulf, down around Abu Musa Island and the Straits of Hormuz. Of course, the American hostages were still being held in Teheran, and no one was sure how that was all going to work out. Our normal flight schedule would be five hours in the morning and then five hours in the late afternoon. We had three pilots; Bill Gibson and I were both Helicopter Aircraft Commanders, and Rob Scott was the duty copilot. We would launch for two ½ hours and then return for fuel, then launch for another two ½ hours. The aircraft would then land after five hours in the air. The maintenance folks would service the aircraft, perform any minor repairs, and do a daily turnaround inspection. It was the same routine in the evening.

Politics in the gulf tended to affect our missions. The hotter the rhetoric between both sides, the closer we had to be to the ship. Commander Middle East Force (COMIDEASTFOR) would set those limitations. In addition to the P-3 and the Iranian warships, we had to pay attention to the fighters based in Bandar Abbas. As mentioned before, PETERSON spent the majority of its time near the Straits of Hormuz and close to Abu Musa Island. The Iranian fighters from Banda Abbas could take off and be on station over us within a couple of minutes. We had a term "Hot Pipe" that we used when the fighters would launch out of Bandar Abbas. If we were flying, we would get low and slow and interpose the PETERSON between us and the potential Iranian threat. If we were away from the PETERSON, we would try to find a tanker or some commercial ship traffic to cozy up to. One day around Christmas, we were out flying and climbed to about seven thousand feet, where it was nice and cool. There were Christmas carols being played over the ICS, and our thoughts were of home when the ship broadcast "hot pipe, hot pipe." We dumped the collective, stepped on the ball, and pointed the nose down at the water. We beat the fighter down to the deck and cozied up to an oil tanker with time to spare. But we did not climb up to a higher altitude to listen to Christmas carols after that.

The first time the CO of the ship tried to dictate how we would do our maintenance on the aircraft was when we were having a problem with our tail rotor drive shaft. The tail rotor drive shaft connects the combining gearbox to the intermediate gearbox and then to the tail rotor gearbox. The shaft that connects the gearboxes is comprised of several sections. The sections are connected by bearings that are self-centering. The shaft and the bearing are supported by pillow block brackets with temperature-sensitive tape. If the tape turned black, then the temperature of the bearing was too high. We had been turning one of the tapes black and had gone through all our maintenance procedures to solve it. BG briefed the CO on what we were doing, and the CO said, "I have some guys that can weld that shaft into one piece, and then you won't have that problem." The OINC explained that the reason there were individual sections rather than one shaft was that there had to be a certain amount of give in the couplings to keep them from destroying themselves. We eventually had a technical representative come to the ship to help us with the problem. We replaced one of the pillow blocks and still were turning the temperature tapes black. The Anti-Submarine Warfare Technician

(AX) on the det came up one day and said, "the old pillow block supports have probably been on this airplane since it was built. Do you suppose after all these years, there might have been some torque on the tail fuselage, and the position of the pillow block had slightly changed? What if we put a shim under the pillow block?" Now the AX is the guy who fixes the ASW gear; he does not do engines and airframes or mechanical stuff like that. We looked at one another and said, what do we have to lose? We manufactured the shim, installed it, and the black temperature tapes went away.

Our next modification issue with the CO was more serious and wound up getting BG hammered by the squadron.

Dhows, the small boats that were commonplace in the gulf, could pose a threat to the ship. But what really worried the CO was the mines in the water. The Iranians were known to have this mining capability, as the U.S. Navy would painfully learn years later in the gulf. The CO wanted to know if we could install machine guns on the aircraft. That way, we could use the machine guns in the H-2 to blow up the mines when we found them without endangering the ship. The issue of arming Navy helicopters in general, and especially the H-2, had been a source of heated debate for years. H-2s had been used for Search and Rescue in Vietnam, and one of the few Medals of Honor given to a Naval Aviator was for a rescue made using an H-2 in North Vietnam. But we were not trained, equipped, or authorized to put guns on the aircraft.

When the CO first raised the issue with the OINC, Bill Gibson tried to shut down the CO's idea completely. BG gave the CO the correct answer; we were not trained, equipped, or authorized to do it. Over a few days, it became obvious to BG that the CO would not let up. AMHC Dusty Hoffman and I put our heads together and produced a plan on how to jury rig the guns and how we would coordinate the firing of the guns by the Aircrewmen in the back of the aircraft.

ARMED H-2 (M-60 IN THE DOORWAY)

L-R LT COYNE, LCDR GIBSON

The OINC went back to the CO and said it was possible to mount and operate the M-60 machine guns but that we would need permission from the Squadron and AIRLANT to do so on a permanent basis. After more pressure from the CO, we decided to do a test flight as our proof of concept. We would use those results to ask the Squadron and AIRLANT for a waiver to modify the aircraft. I must confess I got caught up in the making it happen part of this whole evolution and did not really give enough thought to how the higher authority would react. BG could not get the CO to release a message back to the Squadron or AIRLANT to authorize the modification. Because of the operational schedule, the only time we left our patrol area was to get fuel and so we did not have a chance to get to a phone so that we could call the squadron and let them know what was going on.

The CO of the ship also would not allow BG to send personal for (P4) to Skipper Clay back in Norfolk to tell him what was going on. Finally, we decided to use the detachment administrative message to the squadron to make the folks in Norfolk aware of what was happening. Every ten days, we sent a message to the squadron that told them our flight hours, fuel expenditures, and operational information and contained a section where we would request administrative assistance. This section normally had information like advancement exam recommendations, pay, and earnings stuff etc. AMHC Hoffman had been a former aircrewman in the EC-121 Super Constellation. So, in the administrative section of the detachment message, we requested flight orders for AMHC Hoffman and listed his duties as a left-door gunner.

In about forty-eight hours, the universe clouded up and rained all over us. Our first indication was an announcement over the ship's 1MC. "LCDR Gibson – bridge." About twenty minutes later, BG returned, minus his lips, which the CO had ripped off. BG said the thing the CO was most angry about was that he could not figure out how the squadron and AIRLANT had found out about the unauthorized modification. Apparently, Commander Middle East Force (COMIDEASTFOR), a Navy One Star, had personally talked with the CO of the ship and told him there would be no aircraft modifications without prior approval and where did a surface ship CO get off thinking he could modify an AIRLANT aircraft on his own authority? The det was in the doghouse for the next couple of weeks with the ship, but our outlook brightened when we found out that we would be getting a couple of days ashore in Bahrain. Little did we realize that the worst was yet to come professionally over this whole incident.

Bahrain looked surprisingly good as we pulled in. We all looked forward to getting off the ship, exploring Manama, and having a few cocktails. But first, we needed to call home and check in with the squadron. BG and I went to the Navy administrative office in Bahrain to contact the squadron. BG called and asked to speak with Skipper Clay. I could tell right away that the conversation was not going well. BG looked like someone had punched him in the gut. It was a very short conversation, and the Skipper did not give BG much of an opportunity to explain why we had done what we did. In fact, BG listened a lot and never did have a chance to explain what had happened to the Skipper. I believe the first words out of BG's mouth after he hung up were something like, "there goes my naval career!" The Skipper said there was no excuse for what we had done. CDR Clay said that BG should have been relieved by the ship CO for refusing an order to modify the aircraft and not giving in to the CO. The Skipper said he would have defended BG if he had been relieved by the CO of the ship but could not accept us modifying the aircraft without permission. He said we all knew that we should not have done it. There was no mention of all the hours we had flown, our fully mission-capable rate, or an appreciation for the missions we were flying in one of the most dangerous areas of the world. We left the administrative building and adjourned to the nearest bar, where we celebrated the end of our glorious naval careers.

We were not popular with the ship or the squadron at that point. At times like that, there is often an incident that galvanizes the group and unites them behind a common purpose. In the gulf, the weekends were observed on Friday and Saturday. So, in effect, Saturday became Sunday, and the ship observed what is called Holiday Routine. You could sleep in later; there were no scheduled operations, and folks could take a break. What rankled the det was that we were required to move and spot the aircraft on the flight deck on Saturday so that the ships' company could play volleyball in the aircraft hangar. It seems like a small thing, but it became more of an issue over time. One Saturday, I was in my rack observing the holiday routine when I heard the ship's 1MC announce, "LCDR

Gibson – bridge!" Knowing that something bad must have happened, I went to the hangar. A bunch of the ship's company folks was gathered around our crew leader AMHC Hoffman. Chief Hoffman reminded me of a wire-haired terrier due to his physical stature, beard, and aggressive attitude when he thought he was right. The Chief was in the face of the ship's Master Chief and the XO of the ship about something. When BG arrived in the hangar, he told the Chief and me to go to the det office. Once we closed the door, BG looked at the Chief and asked, "did you really throw the volleyball over the side?" The Chief said, "Yes sir, I did, and I will keep throwing those mother******'s over the side if they come near me!" The AMHC had been out on the flight deck, and the volleyball players kept knocking the ball out onto the flight deck and then laughing and telling the Chief to get it for them. It happened a couple of times before the Chief deep-sixed the volleyball by throwing it over the side of the ship into the gulf. BG told the Chief there would be no more throwing volley balls over the side and to get out. After the door closed, BG and I looked at one another and burst out laughing. The ship ordered a lot of volleyballs, posted a guard when the Chief was around, and there were no more swimming volleyballs. But morale in the detachment improved significantly.

We were sitting near the straits of Hormuz on the day the Iranians released the hostages. The entire mood of the ship changed overnight with their release and the inauguration of President Reagan. I certainly felt that under the Carter Administration, we would have to accept the first blow from the Iranians before we would be allowed to respond. With Reagan in the White House, I felt that we would not be required to take anything from the Iranians before forcefully responding.

The months went by, and we focused on flying and maintaining the aircraft. We flew quite a bit, and it became commonplace for me to fly four hops and ten hours a day. If you flew more than one hundred hours in a month, there was a requirement to be seen by a flight surgeon. There were not any flight surgeons on PETERSON or in the southern Persian Gulf, so we got together with the Independent Duty Corpsman on the ship and put together a message explaining our situation. The Corpsman took all my vital signs, and we sent off a message to the nearest carrier requesting a waiver by the senior flight surgeon if I exceeded one hundred flight hours for the month. A couple of days later, the message came back with a blanket clearance that said that I was good to go if there was no change in my vital signs. I wound up flying 88 hours that month and 106 hours the following month. During that two-month period, the detachment flew over 278 hours with a single aircraft.

It was getting towards the end of the cruise, and we pulled into Bahrain before leaving for home. We called the squadron, and BG talked with the Skipper about how things had been going. With all the flight time that we had accumulated, the detachment's ranking in the squadron was good. The Skipper and BG talked for a few minutes when he motioned for me to take the phone. The Skipper was still on the line and said he had some good news for me. "You are going to have your own detachment. After you get home in March, you will work up detachment Four during April and May. You will deploy on USS CONNOLE FF-1056 out of Newport, Rhode Island, in June."

I did not know whether to laugh or cry. I had been working towards being an OINC over the previous three years, but I figured if it were going to happen, it would not happen until the end of my tour in HSL-34. Turns out the OINC, who was scheduled to go on the detachment, developed a medical problem, and the squadron did not have an LCDR that they could get ready in the period available. The Skipper told me he had spoken with my wife, Linda, and let her know that I did not have a say in whether or not I was going to take the job. BG was on back-to-back long cruises, so he was very sympathetic about what I would be going through.

The ship returned to our patrol area, and we continued our previous operational tempo. The intelligence folks had been searching for a radar site in Iran for quite a while with no success. We were out flying one night, paralleling the coast of Iran in international waters. I was bored, so I reached up and turned off all our external lights. No reason. I just wanted to do it. A few seconds later, our ALR-66 threat emitter lit up because an Iranian surface search radar was tracking us. I turned the lights on, and the Iranian radar stopped interrogating us. I called PETERSON and asked if they had seen the same thing. I told them to call the intelligence spooks on the ship and tell them to get ready to triangulate on the radar. When the ship called back to say they were ready, I turned off the lights. Once again, the Iranian radar lit us up. I kept the lights off as we cruised down the coast for a couple of minutes. Eventually, the Iranians turned their radar off, but the intelligence folks had the information they had been looking for.

During the last couple of months in the gulf, I experienced a series of incidents that got my attention and reinforced the requirement to keep focused when flying the aircraft. The incidents included a single-engine landing, a shipboard landing at night with no lights, and the night I almost hit the hangar.

We were out flying around one afternoon when I heard an engine accelerating. I looked at the instruments and saw that the power was increasing on one of the engines even though we had not increased the power requirement. It was a classic flex drive shaft failure. The flex drive shaft automatically sent a signal to the engine fuel control to increase or decrease power on the engine depending on the position of the collective. When the flex drive shaft failed, it signaled the engine to go to maximum power. Unless we increased collective, the engine would overspeed and destroy itself. I had practiced this emergency quite a lot in the H-2 simulator and during emergency procedure flights in the aircraft. I announced in my best Chuck Yeager, former FRS instructor voice, "flex drive shaft failure on #1 engine, I'm increasing power to prevent the overspeed, request you confirm that I have identified the correct engine, and get out the emergency procedures checklist".

Rob got out the checklist, and we proceeded to complete the emergency procedures. What you did was remove the automatic functions of the fuel control by going to emergency throttle or manual control to compensate for the movement of the collective in increasing or decreasing power on the engine. Now the pilot would manually control those inputs. You had to be careful to lead your power adjustments with the trim switch to increase or decrease power requirements and to prevent the engine from over or under-speeding. We had climbed several thousand feet while we went through the checklist and got the engine under control. I called the ship, declared an emergency, and requested an immediate green deck. We had not been flying long, and so the aircraft was heavy with fuel. My brain malfunctioned, and I did not even think about jettisoning the auxiliary fuel tanks or dumping the fuel in the aft tank to get the aircraft as light as possible when making the landing. Those actions would have significantly reduced the weight of the aircraft by over a thousand pounds. I was assuming in my mind that the affected engine would be ok. This would be a normal landing, albeit with one engine in emergency throttle, something that we regularly practiced. The landing and approach went well right up until the point where I was about to cross the deck edge. At that point, I heard the "bad" engine winding down, and the aircraft began to lose altitude and settle toward the deck. I increased power slightly on the "good" engine but still heard the rotor speed begin to decrease. The slight power pull got me over the edge of the flight deck, and I made a no-hover landing. After the shutdown, BG asked me why I did not jettison the excess fuel. I told him I did not even think about it. We subsequently changed that engine. For the rest of my time flying in the H-2 I made it a point to train my pilots on what to do when, not if, an engine failed while flying. The first thing I told them was to get as light as possible before landing by dumping fuel and jettisoning any sonobuoys or ordnance. This type of incident would come up again on my next cruise and get me in trouble with Commander Fleet Air Mediterranean (COMFAIRMED), but that is another story on another cruise.

I did not know that you would ever have to intentionally land an H-2 aboard a Spruance Class Destroyer at night without any shipboard lights, right up until the point I had to do it. We were out on the last flight of the night and got a call from the ship. They had experienced an electrical failure, and they could not get the lights on the flight deck or hangar to work. I returned to the vicinity of the ship and began to discuss our options with BG and the air controller in the ship's combat information center. I had enough gas to get ashore at that point but would shortly reach bingo fuel. I had no desire to divert, declare an emergency, and land ashore at a strange airfield in the middle of the night. I began to consider my situation in trying to land aboard the ship. We would not have the deck lighting or the shipboard glide slope indicator to assist us in making the approach. On the other hand, the moon was bright enough that we could clearly see the flight deck. When bingo time came, I told the ship we would stay and take our chances on the lights coming back on. We hit thirty minutes of fuel left, and the lights were still not on. I asked the ship to set flight quarters and told them that the Landing Signalman Enlisted (LSE) should be the only one on deck when we made the approach. I considered turning on my searchlight during the approach in order to have better visibility of the flight deck, but I did not want to blind the LSE, who would provide lineup and deck clearance information as I was attempting to land the aircraft. There was about a ¾ moon, and the seas were fairly calm, with no noticeable pitch or roll by the ship, which made me feel that I could manage the approach. The approach and landing went fine, but I knew I would never want to try that on a black-ass night when the ship was rocking and rolling.

The night I almost hit the hangar was a textbook example of complacency kills, or in my case, it almost killed - me.

I had already flown ten hours that day, and we were coming in for our final approach of the night. We called the ship and notified them we were inbound, left seat approach, chocks, and chains for shutdown. I was in the right seat, and Rob was in the left, and he was going to make the final landing as he would be on the side of the aircraft closest to the ship's hangar.

Everything was normal as we made our way around the pattern and turned on to the final approach. His lineup, airspeed, and altitude were perfect, and we had a green ball from the glide slope indicator. At about a ¼ of a mile, I called lineup, and the altitude was fine, but the airspeed was a little fast. At about one-eighth of a mile, I told him everything was good but to slow his airspeed and rate of closure to the ship. I was slow in recognizing that the aircraft's rate of closure was too fast as we were approaching the edge of the flight deck. As we were crossing the deck edge, I realized that we were going to hit the hangar. I announced on the ICS, "I have the aircraft." I threw the cyclic hard right to get away from the ship. I was anticipating impact by the main rotor blades on the hangar at any second. When I realized we had missed the hangar, I steadied the aircraft in a hover and then climbed out. After I turned downwind, using my best Chuck Yeager voice, I told Rob on the ICS that he had the aircraft and that he needed to carry a little less airspeed when he crossed the deck edge. He rogered up and took control of the aircraft. It was a good thing it was nighttime because he could not see my entire body shaking in the seat next to him. His next approach was perfect. After we shut down, I told Rob to go ahead and start the paperwork in the hangar, and we would debrief in a minute. The Chief was standing at the foul line, waiting for me after I completed the post-flight. His first words to me were direct and to the point, "You SOB, don't you ever do that to me again. I don't know how you didn't hit the frigging hangar!" I tried the Chuck Yeager thing again and replied, "Chief, it wasn't even that close." He replied, "oh yeah, the entire flight deck crew and fire party all bolted for the bow. Even the hot suit men were running. That's how close it was!" I was much more vigilant for the rest of the cruise, especially at night.

Ships company referred to the PETERSON as "Proud Pete." Members of the detachment called the ship "Proud Paint." The CO was determined the ship would always look like it had just come out of the shipyard. No running rust or areas that needed to be repainted. Whenever we pulled into port, which was not very frequent, the Bosun's Mates would be over the side painting the ship. Even at sea, there was a constant paint schedule that would eventually lead to the ship asking for a member of the detachment to be assigned to their paint party. When they first approached BG, they asked for an airman to be made available on a regular basis to help with the painting. We slow rolled the ship for as long as possible. Finally, I went to BG with the idea that the Chief and I had thought of. BG went to the XO of the ship and said we would provide a man for the paint party, a second-class petty officer. The guy we are assigning is an Aviation Metalsmith and a graduate of the Navy's paint school. He will be able to ensure that the ship surface is properly prepared, the paint properly applied, and there is minimum disruption of the ship's schedule. We never heard anything about providing a painter again.

Then there was the issue of the passageway in front of the detachment's office. The ship said the det was responsible for maintaining it. The Chief tasked the Aircrewmen with making sure the passageway was properly maintained. The AWs generally were responsible for corrosion control in the aircraft's cabin and would assist the other maintainers in their daily duties, but most of their time was taken up with flying. The AWs took on the challenge and went to a whole different level. They repainted the entire passageway, shined the brass until it was like a mirror, and stripped and waxed the deck until it gleamed. The ship came back and said we had used the wrong paint. Our paint guy pointed out it was the right paint. The ship said it was too shiny and that we needed to make it a less glossy color. The AWs sensed their victory and dulled the finish down a bit.

The last example of paint phobia was sitting in the Gulf Stream the day before we pulled into Norfolk so that we could touch up the paint scheme. The ship had added a few knots of speed to our transit so that they would have about 12 hours to drift in the Gulf Stream while we painted. The hardest part was that we were within range of Norfolk and could have flown off.

As we pulled into Norfolk, I was torn in a lot of ways. I hated the thought of leaving Detachment One. They were a great bunch of guys who had worked their tails off. We had succeeded despite the ship. I was comfortable with them and knew that if I were taking over this detachment, then we would just continue to do remarkable things. I was conflicted about wrapping things up as a detachment maintenance officer while anticipating and looking forward to taking over my own detachment. I felt that, somehow, I was cheating on them. I loved working for BG and the Chief and felt that I still had a lot to learn from them. I also had that doubt in the back of my mind about how I would manage to be the Boss.

HOMECOMING HSL-34 DET ONE

AUTHOR AND WIFE LINDA

When we got back to the squadron, I went in to get a debrief on my FITREP from the Skipper CDR Hank Clay. It was a pack FITREP. I said I was confused. "You are giving me a detachment, but I'm not one of the top Lieutenant in the squadron," I asked. He said that was right. I know you can do the job, or I would not have picked you, but you were not one of the best Lieutenant in the squadron during the last reporting period. I have confidence in your ability to operate a detachment at sea. Your performance as OINC will determine where you fall out in the next round of FITREPS. I walked out of the office determined to have the detachment with the most flight time, highest maintenance readiness, and most ASW contact time in the entire squadron.

The time seemed to fly when Detachment One returned. I was trying to balance the professional with the personal. I had indicated to the Skipper back when I talked to him in Bahrain that I would need at least a couple of weeks leave with the family when we returned, and he said OK. Those two weeks seemed like only a moment in time, and as Linda said, "you never really came off cruise mentally. You just shifted from Det 1 to Det 4."

I do not know how we could have squeezed any more activity into the time when I took over the det in mid-March until we left on deployment in June.

In March, I flew with CDR Clay from Norfolk to Newport, Rhode Island, so the Skipper and the CO of the CONNOLE, Commander Sergey Yonov, could meet, and I could look at the ship and get to know the department heads a little bit before we embarked. I had an exceptionally good first impression of the ship. The CONNOLE had a tactical towed array system for use in ASW, and the CO said that having an embarked LAMPS helicopter enhanced the combat capability of his ship. He said he wanted us to fly as much as possible. I was still skeptical, but it was certainly the kind of language that I wanted to hear from a ship CO after coming off PETERSON.

Next, we would go to the Atlantic Underseas Test and Evaluation Center (AUTEC) on Andros Island in the Bahamas to get our anti-submarine warfare qualifications up to speed and to practice dropping Mark-46 torpedoes against a U.S. submarine. This would be my first opportunity to work with the detachment and to fly with the other pilots who were assigned to me.

We flew down to the Bahamas, and our first scheduled period on the range was in the middle of the night. All the officers on the detachment were Lieutenants. Lynn Howard was the Maintenance

Officer, Wayne Lachowicz was Admin, and Dave Babcock was Operations. AD1 Rick Fisher was the crew leader with AMH2 Myers, AW3 Johnson, AD2 Bryon, AEAN Scelza, AX2 Jackson, AE2 Petty, AW2 Collins, AMH3 McDonald, ADAN Dillingham, and AT3 Thompson. Petty Officers Collins and Johnson were our aircrewmen.

On one of our first flights on the range, I was flying with "Junior" Lieutenant Dave Babcock, who was running the tactical problem. We were given an initial datum to begin looking for the submarine. The H-2 at the time had a tactical plotting machine called the PT-429 that had been taken from the fixed-wing P-2V, the P-3Orion's predecessor. It was better than using a Mark VI manual plotting board, but not much. You could switch scales and plot out the drop of your sonobuoys and have a visual representation of what the pattern looked like and where you were in relation to the pattern. The key was to make sure that the scale selected was big enough to ensure that the indicator or "bug" did not run off the screen. I looked down at the screen and saw that the bug was wedged up against the edge of the screen. I asked Junior, "how long has it been that way," He said he did not know. I tried to figure out where we were in relation to the datum and applied my years of experience and computer-like mind to come up with an "SWAG" or scientific wild-assed guess as to where we were and where we needed to go. In the blockhouse, the tactical observers were laughing because we had about flown off the instrumented range and obviously did not know what we were doing. They looked on with amazement as I turned back onto the range and made a beeline right for the submarine's current position. Of course, I did not know the sub was there; I just wanted to get back on the range and get a sonobuoy into the water. "Madman, Madman, Madman! Smoke away," the sensor operator called out. Our magnetic anomaly detector (MAD) found the magnetic disturbance that a submarine makes as it moves through the water. Junior launched a sonobuoy, and I continued in what we called a MAD hunting circle to get us back over the submarine. The sensor operator came up on the ICS, "Madman, Madman, Madman, smoke away" Now we had a confirmed course and speed for the submarine. I continued in the MAD hunting circle and announced that I had achieved attack criteria and intended to drop my weapon the next time the sensor operator confirmed the submarine's position. Junior made sure the torpedo was armed and said he intended to launch the weapon and then another sonobuoy. I rolled out of turn and lined up on the estimated course of the submarine. The senso called out, "Madman, Madman, Madman," Junior announced, weapon away, and I rolled into another MAD hunting circle. About ten seconds went by when the AW said, "Sir, the sonobuoy launch container is sticking out the side of the aircraft" The sonobuoy launch container (SLC) was screwed into the sonobuoy launcher and contained the sonobuoy. If they were not completely seated, they would come loose when the sonobuoy was launched. The problem was that they were on the left side of the aircraft. The same side of the aircraft as the tail rotor. If the SLC came completely out, it could strike the tail rotor with disastrous results. I informed range control of our issue and told them I would be going to a landing spot so that the sensor operator could remove the sonobuoy launch container. We landed without incident, secured the container, and took off to get back into the problem. I checked in with range control to get a new datum and was told, "return to base" I said the aircraft was fine, and we were ready to resume the problem. Again they said, "return to base" I got a little upset and said I needed time on the range to ensure that we had satisfied our training requirement to track and attack the submarine. Range control repeated, "return to base."

After we shut down, I went into range control to find out why the rest of our time on the range had been canceled. The controllers said the submarine was no longer available. I said how am I supposed to get my qualification if we cannot hunt the submarine.

The Wing Observer said, “You have got the qualification” Usually, you must argue about whether you had tracked the sub to attack criteria. Then there was the issue of weapons placement. Did you achieve the optimum drop point for the weapon to be effective against the submarine? I could not figure out why the wing observer was being so generous. Finally, he looked at me and said, “your torpedo hit the sub” There was no longer any reason for us to discuss whether we had met the criteria for tracking and attacking a submarine.

After AUTEC, we participated in a Composite Training Unit Exercise (COMPTUEX) with the USS NIMITZ (CVN-68) and her battlegroup. The detachment had been selected to take a VX-1 SH-2F to sea to evaluate a new tactical navigation system that was being proposed as an upgrade to the entire H-2 fleet. That entailed working off all the maintenance discrepancies on the VX-1 aircraft to get it ready to go to sea while also trying to maintain the assigned aircraft that we would be taking on the cruise. I was not happy that we were doing maintenance on an aircraft that we would not be taking on cruise instead of the one that we would. I told the maintenance folks in the squadron that I needed time to change a few high-time components on our assigned aircraft that would be due for replacement during the cruise if we flew the number of hours I had set as our goal. Our goal was for five hundred flight hours on the cruise, and we wanted to change out any components that would require replacement while deployed. The squadron said there was no way we would fly that much and so they would not allow us to change the components prior to departure. We wound up changing every major component on the aircraft without shore basing and still averaged one hundred hours a month for the cruise.

A couple of days before we were scheduled to depart, the Skipper called me into his office and said I want your aircraft aboard the CONNOLE two days before you are supposed to leave for the

Med. Get up to Newport and fly aboard while they are still pier side. I was torn. I would lose a full day at home with the family. However, I would not have to worry about getting the aircraft from TF Green Airport to the ship on the day of departure because it would already be aboard. There were also all the issues associated with flying aboard a ship when it was tied to the pier. We anticipated the CONNOLE would balk at trying to arrange the fly-on with Newport Port Operations, but they said sure. We flew up, flew on, and settled in for the cruise. It was hard to watch the ship folks say goodbye to their loved ones.

I had arranged with the CO of CONNOLE to fly the aircraft off and shore base at Naval Air Station Bermuda during our transit to the Mediterranean. We anticipated a day of flying from the Air Station before we would depart for the Med. After we landed, I went into Base Operations to confirm our flight schedule and let the Operations folks know that we would be staying on the ship at the other end of the island. They said their helos were down and asked if we would be available if a Search and Rescue situation came up. I said sure. We rented a couple of 25CC motorbikes to get from the Air Station to the ship, which was on the other side of the island. Lynn Howard and I had barely gotten back aboard the ship when I was told the XO wanted to see me. He wanted to know why we had rented the motorbikes. When the ship arrived at the Navy Annex, which is located on the other side of the island from the airfield, the Bermuda Port briefing team recommended that no one should rent a motorbike because most tourists who did end up in an accident or worse. I knew all of that. Linda had worked at U.S. Customs in Bermuda and told me about the couple that was on their honeymoon and rented the bikes. The husband was going home in a casket because he had collided with the coral walls that line most of the roads in Bermuda. I told the XO we obviously did not know about the prohibition but that we would need the bikes to get back to the air station the next day for our scheduled flight operations. He was not happy, but he did let us keep the bikes.

"LUCY GETS AIRBORNE"

HSL-34 DETACHMENT FOUR ONBOARD USS CONNOLE FF-1056

When we arrived at Base Operations the next day, the Operations Duty Officer said they had received a request for a medical evacuation from a commercial ship that was passing by the island. They estimated the ship was about seventy-five miles from the island, and he asked, can you do it? I said sure, and we started our preflight of the aircraft, followed by a briefing. I said if the ship were big enough, we would land, and the aircrewman could get the sick ship's crewman and load him aboard the aircraft. If we could not land, then we would put a crewman on the deck and then hoist the survivor using a stokes litter. Wayne Lachowicz and I launched and had no trouble finding the ship. Doing this mission with an H-1 required radio contact with the ship or a P-3 to provide us with vectors. In this case, we used our surface search radar in the aircraft and soon sighted the ship. The ship had hatches that appeared to be able to take our weight of about 12,000 pounds. We lowered the aircrewman to talk with the ship captain about the survivor's condition and whether we could land. The aircrewman came back on the radio and said the captain had told him the hatch covers would have no trouble supporting our weight. We shot a normal approach and touched down as gingerly as we could just in case the hatch cover would not take the weight. The crewman approached the aircraft and put the survivor in and was just getting into the aircraft when the ship started to roll. There was no non-skid on the hatch cover, so the aircraft started to slide toward the water. We lifted into a hover as the aircrewman was getting in and departed for Bermuda. En route, we contacted Base Operations with an update on the condition of the survivor and asked them where they wanted us to take him, to the air station or to King Edward VII's Hospital. The controller came back with, "you know where King Edward's is located" I said that I had been stationed in Bermuda from 1975 to 1977 and that if they arranged for a fire truck at the landing site, I would take the survivor directly to the hospital. They said go for it, and we got to fly over the island a little bit before we landed at King Edward's. The next morning, we flew aboard CONNOLE and started the cruise in earnest.

I cannot remember the specific incident, but I will remember the moment forever. Something was wrong with the airplane, the mission was sketchy, and the weather was dog crap. But a final decision had to be made on whether to go or not. At that point, all the heads turned, looked at me, and said, "what do you want to do, boss" Up until that moment, I had not ever really experienced the simultaneous euphoria and stark terror that comes with being in charge. As OINC, there is no one on the ship that is going to overrule your decisions on aviation operations. You are the expert. What you say goes. You are the final decision maker. As a detachment maintenance officer, I always felt confident in making recommendations to the OINC, but they were just that, recommendations. The OINC had to make the final decision and accept the responsibility that goes with that decision. Now it was my turn, and I suddenly developed a whole new sense of appreciation for the OINCs for whom I had worked. Especially when there was no clear guidance to go on in the maintenance manuals or whether to refuse an operational mission.

You develop a little bit of that decision-making ability as an aircraft commander, where you have the final say on whether to launch or continue on a particular mission. The general wisdom in those cases is that if you are standing there after completing the flight and providing an explanation about why you did what you did, then you have done the right thing. The alternative is usually a smoking hole in the ground, and everyone gets to second guess what you should have done, but you are not there to provide your rationale. As time went by, I became more comfortable in those situations and less concerned about what others might say or think. When there was some doubt in my mind, I would always seek as much input as I could get before deciding. Especially as I became more senior, I found that sounding out the room always resulted in a better decision than I would have made on my own with no input from others. However, there were those times when I would not seek any additional advice because I knew what the right thing to do was, and I did not feel I needed any additional input.

We pulled into Rota, Spain, for in-chop and had a chance to meet the Destroyer Squadron Commander or DESRON. He seemed like a good guy and was enthusiastic about how the detachment would enhance the ASW capability of the CONNOLE. On our last night in Rota, we all gathered at the club, and copious quantities of alcohol were consumed. Soon we were conducting carrier qualifications on the tables in the bar. This would consist of an individual launching himself to slide on the table and then trying to catch the belt that ran across it with their toes. Later Lynn Howard and I were playing dice games for drinks with the DESRON and the CO of CONNOLE, and they were the ones that were buying the drinks. One of the CONNOLE officers came up and reminded me that the XO of CONNOLE had ordered us all to be back aboard the ship by midnight. I told him to tell the XO that we were with the CO and would ensure that he got back to the ship safely, whatever time that might be. The CO, Lynn, and I closed the bar and then returned to the ship, aiding one another up the brow. Bright and early, there was a pounding on my stateroom door, and CDR Terry Sheffield, the XO, entered and demanded to know why we had not obeyed his curfew order. I told him that we were with the CO and wanted to make sure that he made it back safely to the ship. A variation on the "leave no man behind" theme. The XO did not buy it and said never to disregard an order of his again.

The ship's officers on the CONNOLE were a good group. We had gone through our arguments and manhood issues and were beginning to really gel. CONNOLE's Captain Yonov made sure the ship's company understood that the detachment was to be treated as a part of the ship and that they should cooperate with us. The XO, Commander Terry Sheffield, was a reasonable type who was open to suggestions on the best way to get things done. The Weapons Officer (WEAPS), Lieutenant Commander Gene Uricoli, had a wry sense of humor, and there were no arguments over the torpedo dolly. The Supply Officer (SUPO), Lieutenant John Hoops, was one of those guys you just enjoyed being around. He seemed to think of everything, and his guys could do magic when it came to producing everything from aircraft parts to uniform requirements. Being the SUPO, he was "the Chop" Supply officers are called chops, short for pork chops, allegedly because their shoulder insignia looks like a pork chop. So SUPO was the Chop, and the assistant supply officer was the "Lamb Chop."

I roomed with the Chief Engineer (CHENG), LCDR Paul Cox. He took especially good care of me. Fresh water is priceless aboard Navy ships, and water conservation is a major concern. When the ship cannot produce enough water, then water hours are established. Water use is severely limited, and in extreme cases, it is used only to prepare food, bathe the folks that prepare the food, and wash the aircraft. The aircraft gets washed to prevent corrosion. It is not unusual to see members of the detachment being washed along with the aircraft when water hours are in effect. On the other end of the spectrum, if the ship is producing too much fresh water, then they will pump it over the side during the mid-watch. A Navy shower normally consists of turning on the water momentarily and getting wet, then turning the water off. Then you lather up with soap, and when you are finished, turn the water on again for no more than 30 seconds or so. A shower that is any longer than that is referred to as a "Hollywood Shower" On CONNOLE, when they were going to pump fresh water over the side on the mid-watch, the door to the room I shared with the CHENG would open, and he would whisper, "Hollywood" I would bolt for the head and a luxurious ninety-second shower.

One day I remarked that all the frigid air in our stateroom seemed to pool at his level in the lower bunk, and the room was warm at my level in the top bunk. About two days later, I came back to the room, and the CHENG pointed to a louver that had been attached to the overhead blower that allowed air to be steered directly at my bunk. The device even had a national stock number on it. The other pilots immediately submitted standard Navy supply requisition forms to have similar devices manufactured and installed in their staterooms.

On CONNOLE, we had "Happy Hour" every Friday. Happy hour was a tradition in the Navy, especially with aviation squadrons. Ashore you would go to the club, have a few drinks, play games, and tell sea stories. The sea stories started with phrases such as, I'll never do that again." Or "there I was when suddenly…. fill in the blank." You learned a lot about the Navy and your aircraft during those sessions. I had never heard of a ship doing it at sea, but we did it on the CONNOLE. We could not have alcohol at sea, but you gathered in the wardroom to play cards, dice, shoot darts, etc. The CO even allowed us to wear hats, which you never did in the wardroom. Most of it was because of the CHOP. He would produce diabolical games and was creative in having different hors' oeuvres. We pulled into Barcelona, and Captain Yonov said we were going to have dining in, a formal dinner, at the Yacht Club. The uniform would be Service Dress Blue Yankee, or we referred to it as Yachting Attire. The dark blue trousers and black shoes we normally wore with our service dress blues were replaced with white trousers and white shoes. The dining-in was a terrific way to bring the wardroom together and was an enormous success.

SERVICE DRESS BLUE "YANKEE" IN BARCELONA

L-R LT BABCOCK, LT HOWARD, LT LACHOWICZ, LT COYNE

Over the next few months, the ship established a great reputation for ASW. I was amazed that we found any submarines at all. We operated at the periphery of the battle group. The P-3s would handle any sub-contacts distant from the battle group. As a sub got closer, the S-3s off the carrier would take over. If the sub continued toward the carrier, the H-3s would take over, using their dipping sonar. Then there was always a U.S. nuclear submarine around that really knew where the Soviet sub was and was tailing it. I used to joke that the ASW commander in the Med would say, "where's the least likely spot for a Soviet submarine to be? Let's send CONNOLE there!" There were occasions when these criteria did not apply. In fact, twice the initial contact on a Soviet sub came from detachment members working on the flight deck.

One of the detachment members happened to look over the side one day and saw a periscope about a ½ mile from the ship. At the time, the carrier was doing an underway replenishment a couple of miles away. We were airborne, but the aircraft was not configured for ASW. We did not have any sonobuoys onboard. En route back to the ship, we trailed our MAD gear and proceeded to the datum that the ship had developed based on our crewmembers' visual on the periscope. We flew to the position and began an expanding search using the MAD. After about ten minutes, the aircrewman said, "Madman, Madman, Madman, smoke away." Just like at AUTEC, I rolled into a MAD hunting circle. We continued to get positive hits on the MAD and continued dropping smokes. Meanwhile, the carrier had dispatched some S-3s and a couple of H-3s. The airspace was getting crowded, and we were out of smokes and low on gas, so I called the CONNOLE and requested flight quarters so we could fuel and load more smokes and sonobuoys. By the time we launched again, the carrier assets had lost contact with the submarine. I knew where the submarine was not, anywhere near the S-3/H-3 furball. I figured that once the H-3s and S-3s started banging on him using active tactics, the submarine had gone deep and fast to get under the surface units and come up on the other side of the battlegroup. So, we talked to the ship and asked for a new datum based on our theory of where the sub might have gone to. We flew to the new datum and put in a couple of sonobuoys. There was no contact, so we expanded our original pattern. About an hour later, as we were continuing to expand off the original datum, the aircrewman sounded, "Madman, Madman, Madman, smoke away." I rolled into the mad hunting circle, where we continued to get in contact. While we were in the MAD pattern, I heard the H-3s and S-3s were inbound. I had had enough for the day, so we gave them the sub's approximate position and called it quits.

By this time, the detachment was running smoothly, and we had hit our operational stride. We were the only all Lieutenant detachment in the Mediterranean. AD-1 Rick Fisher, the crew leader, had the maintenance folks performing at a high level. That professionalism was about to receive a real-world test.

The United States had been at odds with the government in Libya for quite some time. One of the principal areas of disagreement was over freedom of navigation operations (FONOPS) in international waters, in this case, the Gulf of Sidra. The Libyans said that these were territorial waters and that the U.S. could not operate there. They established what they called a line of death and threatened to attack any U.S. Navy units that crossed that line. The U.S. has had similar disagreements with other countries, including our allies, about what constitutes international waters or the right of innocent passage. The innocent passage is when a ship must enter another nation's territorial waters to get from point A to point B. In the Libyan's case, the Navy would periodically engage in what we called "spinner operations." Navy ships and aircraft would operate south of the "line of death." Normally these operations would "spin up" the Libyans, who would respond with their ships or aircraft. We would have a chance to see their tactics and practice ours on them. Depending on the political situation, these encounters could be routine or tense. In our case, it would be the latter.

Our part in the spinner operations would be to conduct anti-ship surveillance and targeting (ASST) and anti-submarine warfare (ASW). We would assist in keeping track of all surface and sub-surface contacts that the Libyans would send out. The Libyan order of battle had some surface platforms and diesel submarines. We would provide targeting information for the surface platforms. Our ASW loadout would be fifteen active sonobuoys and one MK-46 torpedo. WEPS and I got into a manhood issue over a practice loading of the torpedo. WEPS wanted us to do the loading with a practice torpedo. I wanted to use what we called a 'war shot' or regular MK-46. He was worried that we would somehow damage the torpedo. My response was that if I was going to screw up the loading of a war shot, I wanted to do it in practice and not when we might have to do the real thing. After some back and forth, he finally agreed, and we practiced several loads with the war shot. I also had the detachment practice the removal of one of our auxiliary fuel tanks while simultaneously loading the sonobuoy launcher and the active buoys. Our worst-case scenario was to be out on an ASST mission and told to switch over to ASW. For the ASST mission, we would have maximum gas, no sonobuoys or weapons. So, we practiced the worst possible situation. Having to land, shut down, and simultaneously; load fifteen sonobuoys on the left side of the aircraft; drop the auxiliary fuel tank on the right side of the aircraft; load the torpedo on the right side; and refuel the aircraft from the right side. We got good at it, and I was confident if called upon, we could do it for real.

We were out flying around when the F-14s of the NIMITZ splashed the two Libyan fighters. The ship recalled us, and we landed. While the circus of arming the aircraft on the flight deck was going on, I went to the Combat Information Center (CIC). In CIC, we were briefed on where the U.S. units were and where the Libyans were and told that any submarine we encountered was to be considered hostile and immediately engaged. I did not question that at the time. Later I learned that there were other submarines around that would not have taken kindly to us launching a torpedo at them.

After we got airborne, we were asked if we could perform a Search and Rescue mission. When I asked who we would be rescuing, the ship replied, "the Libyan pilots." The Libyans did not have the capability to rescue their own pilots. I said that we were not configured for a SAR mission and, at best, we could lower the hoist and see if the Libyan could get into the rescue harness, but we would not take them into the aircraft. They would dangle on the end of the hoist until we reached the ship. Any funny movement, and I would tell the crewman to cut the rescue hoist cable, and we would let the SOB drown. Fortunately, the Libyans were located and rescued without our assistance. That evening there was cake in the wardroom, and everyone was proud of how the NIMITZ Battlegroup had responded.

CAPTAIN YONOV AND LT COYNE AT THE CEREMONY IN CONNOLE WARDROOM

One of our most important missions was the maintenance of morale onboard the CONNOLE. We did this by aggressively searching, identifying, and exchanging the latest movies with other Navy ships. I mentioned earlier that CONNOLE was a very unusual ship. The CO was a regular every night for movies in the wardroom, along with the XO and some of the junior officers. This had not been the case on other Navy ships that I had been on.

The surface community tended to frown on attendance at movies as a sign that an officer did not have enough to do or they were shirking their duties. In those days, movies were shown on regular projectors. Aviators displayed an unusual ability to be able to change reels while the movie was still running with minimum interruption. So, the aviators assumed the duties of scheduling and running the movies aboard the ship. On CONNOLE, we also used to bet on the movies. Depending on the genre, we would bet on how many people would be killed during the movie or how many breasts would be exposed during the movie. The CO was the arbiter on the final number. The movie was halted one evening when there was a scene in a mortuary. Captain Yonov had to rule on whether the dead in the mortuary should be counted because we had not seen them actually being killed. Then we had to go back and count all the bodies in the mortuary. That night the highest number bid won the pot. Even the officer of the deck would call down on the captain's growler or sound-powered phone to ensure their bets were placed.

The only time I ever remember the movie being interrupted was one night when the CO's growler went off. He was talking with the officer of the deck. The CO asked the OOD to repeat the information. Captain Yonov dropped the phone and rushed out of the wardroom. About a minute later, we heard a series of blasts on the ship's horn and could feel the ship maneuvering. We all bolted for the flight deck and arrived to see a "Grande Maru," a huge oil tanker going down our side about ½ mile away. Ten minutes later, the CO returned, the movie resumed, and that OOD never drove the ship again.

Whenever there was another U.S. ship nearby, the CO would call them up on the radio and see if they had any movies they wanted to trade. If so, we would fly on over and get them. Depending on the type of ship, we would do touch and go's so that their flight deck crew could receive proficiency pay or land and take on fuel so that the flight deck crew could practice their refueling operations. Eventually, we would load the movies that the CO wanted to exchange before we went out on a mission. That way, we did not have to return to pick them up.

One day I got the call to report to combat immediately. When I arrived, the Tactical Action Officer (TAO) told me that our aircraft had called and wanted to talk to me. The boys had been flying around fat, dumb, and happy when one of their engines decided to quit. Flat out stopped running with no warning. My previous experience in the Persian Gulf kicked in, and I asked them about their weapons/fuel status. They had a full load of sonobuoys, and their auxiliary fuel tanks were empty. I told them I wanted them to get as light as possible prior to coming aboard the ship. Lynn Howard was the aircraft commander, and he asked, do you really want me to jettison the sonobuoys? I said, "Yes"! Once the ship was ready at flight quarters, I told them to jettison all the sonobuoys and dump the twelve hundred pounds of gas in their aft fuel tank. This would give them about six hundred pounds of gas to shoot their approach, which would be more than enough.

They were on their downwind leg when I told them to go ahead and jettison the fuel and the sonobuoys. The sonobuoys started coming out the side of the aircraft, and then the aircraft disappeared. I thought it had exploded. Turns out, when you dump twelve hundred pounds of gas, it creates a cloud that hides the aircraft from where I was standing. Lynn Howard, Dave Babcock, and Art Collins did an outstanding job on their single-engine approach and no hover landing. We did the incident report and sent out all the appropriate messages. About a week later, we got a blast from the ASW folks in the Med demanding to know why we had used fifteen sonobuoys without reporting any submarine contacts. The CO of CONNOLE would not let me send my original message and suggested that I tone it down as follows. "Regret use of sonobuoys without generating any contact. Use of sonobuoys deemed expedient to save aircrew and aircraft during single-engine failure. OINC sends." Later Lynn, Junior, and Art were named AIRLANT Pros of the week for the way they managed the emergency.

We were back down by the line of death flying on a Sunday when we spotted a surface ship and ran over to identify it. We normally "rigged" a ship by providing a verbal description of it. In cases where the ship was of special interest, we would carry a camera, and the aircrewman would shoot pictures. We passed on the information to the CONNOLE and continued to fly around for about the next hour and fifteen minutes. Eventually, CONNOLE called and wanted to know where that contact was. We told them what the course and speed were and an estimate of its position. Could we find it again? Sure, right after we get some gas. Turns out it was a ship the intelligence guys in the Battle Group were interested in, but nobody at the higher level was paying attention when the CONNOLE passed the information.

HSL-34 DET FOUR ONBOARD USS CONNOLE FF-1056

THE AUTHOR IS SEATED IN THE LAWN CHAIR HOLDING SUNTAN LOTION & BOOM BOX

Another time we were out flying right along the line of death and saw a submarine on the surface. It was about 5-10 miles inside the line of death. We were not supposed to go south of the line without permission. We told the Combat Information Center (CIC) on the CONNOLE the subclass, course, and estimated speed and that there was a number on the sail, but we could not read it from where we were. CIC asked, can you give us the number? I repeated the sub was inside the line of death, and we could not see the number clearly. Over the next 15-20 minutes, CIC would call back for more information. Are you sure of their course and speed? Can you get a better angle to see the number? It was obvious they were getting pressure from higher authority to give them more information. Finally, I turned off our transponder, descended to one hundred feet above the water, and set course to intercept the sub. The ship started calling me, and I ignored them. Then the battlegroup called on UHF guard. We intercepted the sub, flew over the top, and then proceeded back to the north. After we recrossed the line of death, I turned on the transponder, called the ship, and gave them all the information that they had been requesting. I did not know what would happen next. We had crossed the line of death without permission, turned our transponder off, and failed to respond to radio calls. No one on the ship or the battlegroup ever said a word about it.

It was the end of the cruise, and we were all thinking about going home. We had one more port call to make in Tunisia for four days. WEPS was the senior watch officer on the ship, and I said the officers' on the detachment would take the shore patrol assignment for the in-port period in Tunis if he agreed. The Shore Patrol Officer managed the liaison between the local authorities and the ship. They would mediate any incidents that occurred during the port visit between sailors and the local merchants. The officer would spend the night at a hotel in town so that they were immediately available if anything should happen. I explained to the other pilots that if we took the entire period, the room would be ours, and we would have a place to crash while on liberty at the expense of the U.S. Navy. Everything was set to go when Anwar Sadat was assassinated in Egypt. The port call was canceled, and we joined other Navy units off the coast of Egypt. No one knew what was going to happen, so we could not even guess how long we would be there and what effect this would have on us going home. We drilled around for about two weeks before the National Command Authority decided the CONNOLE should return home.

When I returned to the squadron, they were not sure what to do with me. All the department head and assistant department head jobs were being filled with Lieutenant Commanders. There were no slots in Operations or Maintenance. So, I became the Assistant Safety Officer and the Naval Air Training and Operating Procedures Standardization (NATOPS) Officer. I was not thrilled about the Assistant Safety Officer job. In the naval aviation hierarchy, the safety officer is not one of the more desirable positions. It is a vitally important job, but the nature of the position is that you are constantly telling the CO/XO about things that are not going well and that could result in an accident or injury, things that could end their careers. That tends to make you not so popular with the command leadership. The Safety Officer agreed that I should concentrate on the NATOPS job and help as the Ground Safety Officer.

The NATOPS job was challenging. We rewrote the NATOPS training syllabus and produced new open and closed-book examinations. There was a helicopter hanging from the ceiling over my desk with the inscription, “A Day without autorotation’s is like a day without sunshine.” As a NATOPS officer, I was supposed to be the squadron expert on the aircraft systems, normal and emergency procedures, limitations, etc. I did not have as much overall flight time experience in the H-2 as the CO/XO and Department Heads, but I had more time in the aircraft over the past year than anyone in the squadron. The fact that I had just done two deployments and flown hundreds of hours the previous year gave me the credibility to address the squadron about what the book (NATOPS) said. I spent a lot of time prepping the pilots and aircrew for our upcoming squadron NATOPS evaluation, which we later passed with flying colors.

Then I learned that I would be going back to sea with the folks on Detachment One. They were working up for another deployment, and none of the pilots on the detachment had become a maintenance functional check pilot. The OINC had previously been a functional check pilot but had not completed the syllabus for requalification. So, I would go along as the fifth pilot and fill in as needed tactically while doing any necessary maintenance check flights. It was a strange position to be in. The new OINC was finding his way after coming from a shore assignment, and I did not want to intrude on his attempts to mold the detachment to do things his way. But, I had previously cruised with Chief Hoffman, and based on our experience together, he knew that I did not believe in some of the things the OINC wanted to do with the detachment.

One thing I did agree with was the OINC’s determination to work the det up gradually. We were in company with another ship that had a detachment from our sister squadron embarked. Although they were in the same phase of their workup cycle as Detachment One, they were flying significantly more. All the squadrons on the East Coast competed to see who could fly the most flight hours while ashore and deployed. This was what I called the “Great East

Coast Flight Hour War." I had willingly done my part to try and ensure that HSL-34 was at the top of that list. But Detachment One was still in the process of learning about one another and was not flying as much as the HSL-32 detachment.

The weather had been bad, and the OINC had called it quits on flight operations for the evening. In my rack, I heard "emergency flight quarters" broadcast over the ship's communication system. Up in the hangar, the OINC, Chief Hoffman, and I tried to figure out what was going on. The OINC said he was going to combat and asked me to get the aircraft spotted for launch. Chief Hoffman and I were in the process of organizing the aircraft movement. I walked to the door leading out onto the flight deck and opened it. I could not see twenty feet in the fog. I told the Chief to hold on to spotting the aircraft while I was going to go to the Combat Information Center (CIC) and find the OINC. CIC was loud as everyone was trying to get information about the HSL-32 aircraft that had gone into the water. The folks in CIC were asking the OINC about how long it would take to get our aircraft in the air. I tried to get the OINC's attention, but he was focused on giving the ship's CO an idea of how long it would take. Again, I tried to get his attention so that I could talk with him privately. Finally, he looked at me and loudly said, "what do you want?" I said, "if you try to launch with the current weather conditions, the only thing you will do is put four more people in the water! The weather is 0/0. I cannot see the circle on the flight deck from the hangar. We have to wait for the fog to lift before we can launch." The OINC paused for a moment and said he wanted to go evaluate the weather for himself. About five minutes later, he returned and said we would have to wait for the weather to lift before we could launch. About that time, the other ship reported they had the survivors in sight and would pick them up with their motor whaleboat.

After we got back from the workups, the Skipper called me into his office and asked if I would consider taking a detachment back out for special operations. We were expecting our third child, and it was unclear whether the detachment would get back in time for our daughter's birth. I told the Skipper I would go if he ordered me to but that I would not volunteer. There was a little frustration in my voice as I had done two long cruises in the squadron and had just returned from a workup period for another detachment. "Seems like there ought to be somebody else that can take a detachment to sea," I said. The Skipper chose another OINC to do the special operations. If I had gone on the special operations mission, I would not have been back in time for the birth of my daughter Erin.

It was time for me to contact the detailer and arrange for my next set of orders which would be to shore duty. I drove up to Washington, D.C., from Norfolk and sat down with the detailer. He outlined a variety of jobs. None of them seemed particularly exciting. The obvious choice was to return to the FRS in HSL-30 as an instructor. But I needed to get out of the cockpit. I was getting a little complacent in my flying, and that could kill me. While pausing to collect my thoughts, I asked, "do you have my latest FITREP?" He said he did not. I passed him a copy of my latest fitness report, the one I had received after returning from my second consecutive cruise in the squadron.

The detailer finished reading the FITREP, opened a desk drawer, took out a sheet of paper, and started asking me about a list of jobs that were much more appealing and career-enhancing than the ones he had previously been talking about. The FITREP I handed him said that I was "#1 of 12 Lieutenants in this command. As the only Lieutenant Officer in Charge to make an extended deployment with a detachment in the past year, LT Coyne has demonstrated the leadership necessary to take an inexperienced crew to sea and return home with more flight time, mission time, and ASW contact time than the rest of the dets combined. As an aviator, LT Coyne has over six hundred accident-free flight hours in the preceding year with ninety percent of that time embarked." I had always heard the Navy had different lists of potential jobs depending on your performance. The detailer offered me a job on the OPNAV staff that had recently been established and worked on special interest correspondence for signature by the Secretary of Defense, Secretary of the Navy, and the Chief of Naval Operations. I decided to go to Washington, D.C., and the Navy Staff.

CHAPTER VII

PUZZLE PALACE

OFFICE OF THE CHIEF OF NAVAL OPERATIONS

GENERAL PLANNING AND PROGRAMMING DIVISION

OP-906E6

JUNE 1982 – FEBRUARY 1985

Washington would check all the boxes that I felt needed to be filled. Going to the Pentagon, also known as the Puzzle Palace, would allow me to see how the shore-based administrative Navy really worked, get some time out of the cockpit, and try to pick up a master's degree. The hours would be long, but I would be home at night, and we would have more time for family activities. The job I had chosen had only been established within the previous six months and was in the Planning and Policy Special Interest Correspondence Section as OP-906E6.

The Secretary of the Navy and the Chief of Naval Operations were getting beat up by Congress over their failure to respond to Congressional Inquiries promptly. Senior Navy leadership decided to set up an office whose mandate was to turn around the answer to any assigned Congressional Inquiry within four days. We had special orange folders that distinguished anything that came out of our office from any other Navy correspondence in Washington. We also had a cover sheet that said whoever received our correspondence and proposed response had ninety-six hours to respond. If they did not respond, concurrence would be assumed. Our chain of command for correspondence approval consisted of OP-090, the Vice Chief of Naval Operations, and finally, the Chief of Naval Operations or the Secretary of the Navy. We also managed special interest correspondence for the signature of the Secretary of Defense when Navy issues were involved.

Coming from a background in broadcast journalism, the idea of putting together a response quickly under pressure was not new to me. When an inquiry came in, we would contact the pertinent OPNAV code to see if there was any earlier correspondence on the issue. If there was, we would use that as the basis for the proposed response, updating it to reflect any changes and considering who in Congress was asking the question. If there was no previous response, you would talk to the action officers in the OPNAV code to see what they thought the answer should be. Then you would take their inputs and draft a proposed response that would be sent to the applicable OPNAV codes to review and approve it. This was done simultaneously and vice sequentially, and it was up to me to decide which changes I would take from the various OPNAV codes to be included in the final response. The tough proposed responses were those with no OPNAV or Navy position.

When an issue was assigned, and there was no previous response to base the answer on, the action officers on the OPNAV staff would wish me good luck and that they were looking forward to seeing what I could produce. The ninety-six-hour turnaround time made for some interesting conversations between frocked Lieutenant Commander Coyne and the Captain Executive Assistants for the more senior Flag Officers on the OPNAV staff. Captain X would call three days after the deadline had passed on a proposed response to say that he had some suggested changes. I would tell the Captain I could not do it. Why not? He would ask.

"You will have to call the CNO or VCNO's office because that's who has the proposed response," I would reply. Captain X would say, "You sent that letter forward without my Admiral chopping on it. Who gave you that authority?" I referred him to the cover letter that said concurrence would be assumed unless a response was received.

Then Captain X would ask, "Who is your Captain?" I would tell them, Captain Wesley E. Jordan. They would call Captain Jordan, and the whole argument would start all over again. Word quickly got around the Pentagon that these folks with the orange folders did not wait around beyond ninety-six hours for the approval of the OPNAV Codes. If an action officer had an issue and needed some more time, we had the latitude to extend it but saying you had not had the time to review the proposed correspondence was a non-starter. So here I was, new to the Pentagon, being "wire brushed" by O-6s daily. It was terrifying and exhilarating at the same time.

The whole D.C. experience was different, as I learned while checking in at OPNAV. You reported to a regular office building and found the room number you had been told to find. I found myself sitting in a small, cramped office space across from an O-6 aviator. A Yeoman gave us a series of forms to fill out and then left us on our own. The Captain shook his head and smiled. He looked at me and asked, "Where are you coming from?"

"HSL-34 in Norfolk," I replied.

Then I asked him, "Where are you coming from, Captain?"

He said, "I was the CO of Naval Air Station Roosevelt Roads in Puerto Rico." He laughed again. The pattern in the Navy was that you got proficient at a job or felt comfortable, and then you would be transferred to something that you knew nothing about and were expected to learn at once. It turned out the Captain was the

Executive Assistant or EA to my Admiral, so he would always laugh and ask me how I was adjusting when I needed to brief the Admiral on a proposed response.

The shop was made up of six of us, and I was the junior man as a frocked Lieutenant Commander. There were three surface folks and three aviators. Originally there was another helicopter pilot in the shop. CDR Ray Wickstrom was also a LAMPS guy and especially good at rounding off the rough edges on LCDR Coyne. He taught me the ropes and kept me from doing something really stupid while introducing me to the larger world of naval aviation. He left shortly after I arrived to take command of a LAMPS squadron and his successors were from the jet community. So, I got all the helicopter-related correspondence and the cats and dogs issues that the Commanders in the shop did not want to mess with. Commander Todd Earle was our immediate boss and was great at running interference between us and Captain Jordan. He would review all our proposed responses before we sent them forward and would try to anticipate the things that might set off the Captain or be rejected by the Admiral. He also taught me how to run. He had competed in the Boston Marathon and helped me develop my running workouts. He used to say something to the effect of, you can lie to the IRS, and you can lie to your wife, but you cannot lie to your running log.

The first six months went by quickly, and I settled into my routine. At work by 0700, go to the Pentagon Officer Athletic Club (POAC) around noon time, and leave for home at 1700. The Navy was busy expanding, and I did a lot of answers on shipbuilding programs and homeporting proposals for the six hundred-ship Navy. The Navy had solicited bids for several types of auxiliary ships, and the competition to get the contracts was quite high. We developed a pat response, and whenever a new letter came in, I would call the OPNAV Action Officer to have him give me a verbal run down on what the proposed response would look like before I put my finger to the computer keyboard. We used templates for the letters, and we also had a couple of pages of terms or words that the CNO, SECNAV, and SECDEF did not like. Over time you knew what the CNO, SECNAV, and SECDEF would sign off on. It was always amusing when an EA would call and say his three star would not approve the prepared response unless it had this specific language in it. I would tell the EA that I could not forward a proposed response with that language in it because those higher in the chain of command would reject it. The EA would send his Flag Officer's approved proposed response with forbidden words/phrases in it. I would remove the forbidden words/phrases and send the proposed response up the chain of command.

Our other guidance was that if the proposed response were for the signature of the Secretary of the Navy, John Lehman, it could not be more than one page. There were a few exceptions, but they were rare. The SECNAV believed that if you could not explain an issue on one page, you did not know what you were talking about. I was always uncomfortable with that approach because it assumed that both the writer and the signer knew that to get down to one page, you had to leave a lot of stuff out of the response, some of which might be important. I thought that a Secretary Lehman response on the contents of the bible would consist of: "Old Testament, God is harsh, and everyone dies. New Testament, Jesus saves us, and we all live happily ever after."

The proposed correspondence was accompanied by a staff paper called a Blue Blazer. The Blue Blazer would have all the information that you left out of the letter but was pertinent to understanding the proposed response. It was actually printed on a blue sheet of paper. Normally the Blue Blazer also was no more than a page long. The headings on the blue blazer were Purpose, Background, Discussion, Coordination, and Recommendation.

Two responses I authored were quite different and offered an insight into the Pentagon and politics. They were both concerned with the homeporting of Navy ships. The Navy was pushing to increase the size of the Fleet to six hundred ships. To get Congressional support, the SECNAV was using the possibility of homeporting a ship in certain Congressional Districts. As you can imagine, this generated a lot of interest and a lot of correspondence. The first unique homeporting issue involved a battleship being homeported in New Jersey, and the second involved a couple of frigates in Newport, Rhode Island.

A council member from New York City wrote to the Secretary of Defense and wanted to be assured that there would be no nuclear accidents from a proposed homeporting of a battleship in New Jersey. First of all, Navy policy was to neither confirm nor deny the presence of nuclear weapons aboard a Navy warship. So how do we assure the Councilwoman that nothing bad will happen with nukes, given our policy of non-confirmation? I called the homeporting guys and asked if they had a proposed response. They said no. I asked, what do you think we should say in the response? They said I should talk to the Navy Nukes, and then they would like to see whatever the nukes proposed response was. I called the nukes, explained the situation, and asked what our response should be. They said I should talk with the homeporting guys. They did not have any suggestions or recommendations, but I would like to see whatever I could produce.

Without any guidance, I reverted to what Jack would say to the Councilwoman. The first response I drafted provided the Department of the Navy's non-confirmation policy but affirmed that the Navy was extremely sensitive to the local community's concerns. The Navy would consider those concerns, and I stated that the Navy had never had a nuclear incident. Also, the Navy had a comprehensive plan for homeporting the ship in the New York area.

The homeporting guys loved it, but the nukes said no. I said OK, revise the parts you do not like, and we will send it up the chain of command. They said they would not change anything but would not approve the proposed response as it was drafted. They would not tell me what was objectionable in the response so that I could address their issue. If they proposed a change, this would become their tar baby, and they would be responsible for the letter. Much better to let the LCDR Helicopter Pilot own the letter.

I learned a lot over the next twenty-four hours about what qualified as a nuclear accident, nuclear incident, and whether anyone had ever been injured handling nukes, etc. In short, I could not provide the blanket assurance that the Councilwoman was seeking because there had been incidents and accidents, but not the kind she was talking about. What was an incident or an accident to the Navy Nukes was not what was being implied by the Councilwoman. If you scratched a weapon, that could be an incident or an accident. Therefore, I could not say there had never been an incident or an accident. The Councilwoman was talking about a nuclear device going off for no apparent reason. That had never happened. I was convinced that the Navy, in view of its non-confirmation policy for ships, did have a good plan to prevent any accidents or incidents from happening. I eventually produced a proposed response of about five paragraphs that tried to state the Navy's position.

Essentially, the proposed response said that we know what we are doing, have a good safety record, and the Councilwoman's concerns have been considered in recommending the homeporting of the battleship. When I finished, I sent it out for chop and waited for the proposed changes to come in. No one changed a thing. Not the OPNAV staff, the CNO, or the SECNAV. The letter went to the SECDEF's office and sat for a long time. It was there so long that Captain Jordan stopped bugging me about the status of the letter.

One day I came back from the gym, and there was a letter in my basket. It was from the SECDEF to the Councilwoman and was only about two sentences. As I remember, they were in effect, thanks for your interest in national defense, we know what we are doing, Caspar Weinberger.

The second homeporting letter came from Senator Claiborne Pell of Rhode Island. He had heard the Navy was considering homeporting a couple of frigates in Newport. The Senator reminded Secretary Lehman that the Navy had made a big mistake when it had previously moved ships out of Newport. The Senator said that not enough consideration had been given to the social and economic effects of homeporting Navy ships in Newport and that he was going to look very closely at the Navy's rationale and plans for bringing any ships back to Newport. Things like schools, housing, support facilities, etc. The folks that managed homeporting said to go with the standard response. While no decision had been made yet, the Navy would ensure that proper planning and coordination would occur before that decision. While drafting the response, the OP-40 guys called and said to hold off on the standard response. They were told the SECNAV had decided not to homeport the frigates in Newport. As always, I asked, what should I say? They responded, "We do not know, but we sure look forward to seeing your proposed response!" Back to "What would Jack say in this situation?"

My proposed response thanked Senator Pell for his letter. It said that after considering the Senator's points, the Navy would assign the two frigates to Philadelphia and Charleston instead of Newport. The proposed response went out for chop. Nobody on the Navy staff changed a thing, so I was waiting for a call from the SECNAV's office to come and change the letter because I had been so abrupt with the proposed response. The SECNAV signed the response to Senator Pell with a few minor modifications.

While the SECNAV normally took no prisoners when it came to dealing with Congress, I was really surprised that he did not soften the response. A few months later, while working on a paper for one of my Georgetown classes, I came across some Congressional testimony by then Mr. Lehman testifying before a committee that Senator Pell was a member of. The exchanges between Mr. Lehman and Senator Pell were heated. I thought to myself, payback really is a mother.

After about a year on the job, my Captain called me into his office and told me I was going to be a "bag man" for the Chief of Naval Operations. You know those naval officers you see sitting behind the CNO when he is testifying before Congress? That was going to be me. It was my introduction to how things worked at the highest level in the Navy. Our primary responsibility was to prepare the briefing books that the CNO used to provide information to the Representatives and Senator's questions during his appearances before the Senate and House Armed Services Committees and during the authorization and appropriation hearings. Those same briefing books were also used by the SECNAV. So, we were working for two masters. I would be exposed to and learn about a wide variety of issues affecting the entire Navy instead of just the LAMPS community.

We sent taskers to the OPNAV staff to prepare point papers on the issues that the CNO and SECNAV had identified as important to Congress. These point papers were supposed to be only two pages long.

On one page, you would have the information on the program or issue, the five-year funding plan, which members of Congress had expressed an interest in the program and what their issues were, etc. All the information the SECNAV or CNO needed to know in order to answer any detailed questions from a member of Congress.

On the other page, you would have "clouds" that provided additional information, clarification, or details supporting the point paper. They were called clouds because they had color borders around them in the shape of a cloud and a line that pointed to the text portion of the point paper that would be underlined in the same color. The cloud provided detailed information that was not contained in the point paper. It was like I was back in grade school as I drew the clouds and lines in distinct colors to match them up. After the OPNAV staff provided their input, we would put it all together into a briefing book and then forward the material to the SECNAV and CNO.

The SECNAV and CNO would review the materials and then send them back to us with suggested changes or requests for additional information or clarification. We would take their comments and task the OPNAV staff to provide the information that the SECNAV or CNO had asked for. This iterative process would take place continuously right up until the time the SECNAV and CNO would go to the hill to testify. I was amazed at how much detail the SECNAV and CNO could retain. The CNO made a notation on a point paper that the spending total on the fourth year of the five-year plan for a particular weapons system needed to be corrected. When I went back to the action officer to ask about the funding figures, I could hear him moving papers around before he acknowledged that the CNO was right.

In addition to the Point Papers, we also tried to anticipate the questions from the members of Congress. The Navy Office of Legislative Affairs would provide the initial input, but it was not detailed. They did not want to be wrong on what an individual member of Congress' hot-button issues with the Navy would be. So, we would ask the OPNAV staff that were preparing the point papers what they thought the issues would be. Their questions would be based on correspondence, briefings, or office calls they had made with the members of Congress over the past year.

Over time we produced several categories, questions, dirty questions, dirty rotten questions, and unthinkable questions that would be sent to the OPNAV staff to produce answers for the SECNAV and CNO. The rubber met the road when we went up the hill, and it was Congressman X's or Senator Y's turn to ask questions. Captain Jordan would turn to the briefing book tab that had Congressmen X or Senator Y's name and our question categories and try to anticipate which question the member of Congress would ask so that he could slip the appropriate point paper on that question to the CNO as the member of Congress was pontificating in preparation for the question. The SECNAV had his own guy who would also be simultaneously slipping the answer to the questions forward. We did this whole routine four times before the Senate and House Armed Services Committees for Authorization and Appropriation.

When the transcripts of the testimony came out, we would review them to identify questions for the record. If during the hearing, the Congressional Member's question was particularly contentious, or we screwed up, and the SECNAV or CNO did not have the information to answer it, then they would tell the member of Congress that they would take the Question For the Record (QFR) and respond later. Less formally, the SECNAV or CNO might say to the member of Congress they would be happy to get back to them on that issue. Getting back might consist of a phone call, letter, meeting, briefing, etc. We would identify those things from the transcript, identify the applicable OPNAV point of contact, and then prepare a tasker to tell them what was required and the time frame for the response. The OPNAV staff prepared over two hundred-point papers and responded to five hundred Questions For the Record during that Congressional cycle. All of this made me what I described as the best damn Lieutenant Commander YN (Yeoman) in the Navy who could modify and produce briefing books on demand. It also exploded my aperture on a wide range of Navy issues I would never have been exposed to had I remained in the Fleet.

My overall feeling for the Navy and how things were done was further enhanced when I was selected as the Assistant Recorder for the Reserve Commander Selection Board. I had heard vague whispering about how these boards worked while in the Fleet. By the time it was over, I was more disheartened than positive about the promotion system.

The Board Recorder was responsible to the President of the Selection Board for all the administrative support needed to be able to review an eligible officer's record and make recommendations to the Secretary of the Navy on who should be promoted. The first order of business was for the assistant recorders to review the records of all the eligible officers for the promotion. Our priority was to ensure there were no missing fitness reports in an officer's service record. You had to account for every day the officer had been in the Navy. If there was a gap where a fitness report was missing, we would contact the service member and let them know what we needed. You wanted to ensure that an officer's record contained a current photo. As you went through the records, you also were looking for disciplinary letters or fitness reports that could be considered adverse. If an officer's superior had submitted a letter of recommendation, we highlighted that in an officer's record. We spent a week reviewing all the records and ensuring they were ready for the board.

When the board met, the first order of business was whether to consider letters of recommendation. The board did not have to consider them. As I remember it, they voted to consider the letters.

The room where the board met had a series of screens that could be used to display an officer's service record so that the entire board could see it simultaneously. As assistant recorders, we would put the plastic transparencies that contained an officer's entire career on the projectors that would transfer the image to the screens. This allowed the entire board to simultaneously see and focus on a single record. There was also a different board member assigned to brief the record to the board every time the record was presented.

An officer's record would be considered between one and three times, depending upon when they were selected. Once the briefing was done, the board would vote on each record. The vote consisted of two data points. Recommendation for promotion (Yes or No) and percentage of confidence if promoted. The board members would record their votes using buttons on their chairs. The recorders would tabulate the yes/no and percentage of confidence. We would then rack and stack the records. The board would recommend to the President of the Board that all officers with X number of Yes votes and Y percentage of confidence should be selected. All officers above those points would be considered as being selected, and their records would be set aside. Then the process would be repeated again.

The briefers would highlight parts of an officer's record to draw the board's attention to that particular piece of information. They would also draw on the transparencies. A flower drawn on the record indicated "a late bloomer," an officer whose performance improved as they gained more experience. The board would then decide again on what percentages they wanted to take. Those officers' records would be removed, and they were considered promoted. Those officers remaining to be considered for a third time were in what was called the "crunch zone." This was their last chance at being selected for promotion.

Two unique phenomena that I noticed as the board went on were how quickly the board members would review the service record on the screen and record their votes and the importance of the briefer in "selling" a record to the board. By the third look, we could barely get the record transparencies up on the screen, and the members would already be voting. They had been at this for a week, and their learning curve was complete. They could look at a record very quickly and judge whether that officer should be promoted. The briefing officer was critical during the third look. Over the course of the week, certain members would stand out and have credibility with the other members of the board.

If Captain X had a reputation with the board as an honest broker and he recommended you for promotion, you stood a great chance of making it. If Captain Y was briefing your record and the board did not feel they were credible, you did not stand a chance of being selected. Later in life, I would brief my wardrooms on the necessity of performing well enough so that a board only looked at your record once and selected you for promotion. You did not want to get to the third look, or crunch zone, because your briefer's credibility would significantly impact whether you were selected for promotion. I stressed to the Junior Officers that you had no control over who that would be, but you did over your performance.

An urban legend on selection boards is that an officer was up for his third look, and the briefer said the individual should not be selected. The briefer asked the board members if there were any questions or comments. One of the board members raised his hand and said the officer being considered might not have the best record in the Fleet but that he had flown into Hanoi Harbor to rescue him even though North Vietnamese were about to capture him. The board unanimously voted to promote the officer who, a minute before, was not going to be promoted.

I rounded out my tour in D.C. with a Master of Arts in Government from Georgetown University, being President of the Washington chapter of the Navy Helicopter Association (NHA), and as an escort officer for the Up with People Group that would be performing during President Regan's Inauguration.

Getting an advanced degree in the Navy was career-enhancing. But I did not want to go to Post Graduate School in Monterey, California, to get a degree in systems analysis or anti-submarine warfare. A degree from a respected civilian university is what I was after, and Georgetown had a great National Security Studies program. I did not know if I could get in. My grades were not the best, but I figured the ten years between my application and the last time I attended college might be considered. Also, there was the hope that Georgetown would not want to turn away that G.I. Bill money.

I applied, was accepted, and started classes. My days were filled as I would work 8-10 hours at my OPNAV job and then would go to school at night. Many of the classes were taught in the Pentagon, which cut down on any commute. The thing that also attracted me to the program is that the professors were national security practitioners as opposed to just academics. Many of them had important day jobs in the Department of Defense and then became our instructors at night. The emphasis in the program was on actual events rather than theoretical situations. I remember one of the early discussions involving what was the most important principal factor in national security decision-making, the individual or the job the individual occupied. Over time I came to the idea that in the U.S. process, the quality of the individual determines their influence on national security decision-making rather than the position. The U.S. Army refers to "Iron Majors," the folks who have the most impact in a unit and shape the options that a commander chooses from. However, among our allies, it is frequently the opposite. Rather than the individual, the position the individual occupies determines their influence. The senior officer is expected to be the font of wisdom and does not seek input from those lower in the chain of command.

I was taking two courses a semester. One weekend during the semester, we had a seminar on topics in national security. The seminar would be held over the weekend at a facility in Culpeper, Virginia. You would have assigned reading, and then there would be a series of presentations with different speakers. You had an opportunity at lunch and at the evening dinners on Friday and Saturday to engage the subject matter experts who were giving the presentations. There would be an exam at the end of the semester on the readings for the weekend seminar. My classmates came from across the spectrum of the services and the interagency, and several of them went on to very senior positions in the government.

The program required that you demonstrate language proficiency. I had taken two years of Latin in high school but had avoided any foreign languages in college. As an alternative, you had to demonstrate proficiency in statistics. The only thing worse than my language skills was my mathematical skills. I barely passed a review of basic mathematics courses in college. I signed up for a summer course at George Mason University and got a "C." I had to have at least a "B" to have the credit accepted at Georgetown. No amount of discussion with the professor could get him to change his mind. Despondent, I went to my academic advisor at Georgetown and outlined my problem. He wrote down some information on a paper and said to take this course at American University. The course title was Introduction to Quantitative Analysis in International Relations. It was another course in statistics. I said, "Hey, this didn't go so well at George Mason." He said this course is oriented toward using statistics in national security decision-making, not just statistics for statistics' sake.

I said, "I don't have a lot of time for this. If I don't get a "B" in this course, I won't have time to complete the program before my tour of duty is over." He smiled at me and said, take the course.

The course at American University was not just a classroom lecture on statistics. We had to produce a political theory, devise a data questionnaire, do research for data on our theory, produce a computer program to run the data, punch the data onto computer cards, and then run the data and see if our theory was proven or not. It was much more involved than just attending class, but I learned much more about statistics than I would have in a classroom. This time I got the "A," I needed to fulfill the Georgetown requirement. My family and parents were there when I received my graduate degree from Georgetown in May of 1985. I had not attended my graduation ceremony at Ohio University, so marching at Georgetown was special.

Becoming President of the Washington Chapter of the Navy Helicopter Association (NHA) was another broadening experience. I was never a big "joiner" and had heard about the NHA as a Lieutenant. My impression was that it was more of a West Coast organization. The NHA was based in San Diego, and historically, most of the senior leadership lived there. CDR Ray Wickstrom had taken me to my first meeting shortly after I reported to D.C., and it became a social thing where you could see other helo guys and catch up on the latest in our respective communities. I knew the current President of the Washington Chapter because we had been stationed together in HSL-34. CDR Jim Crawford called me one day and said, "you are the new President of the Washington Chapter of the NHA. "When did we hold the election?" I asked. He said I was elected by unanimous consent. He had chosen me. Once again, it proved to be a valuable experience. We had quarterly lunches, and the President's main job was to get a guest speaker. I was surprised by how receptive senior Navy leadership was to taking time from their busy schedules to come and talk with a bunch of rotor heads. It also exposed me to the more senior leaders in the helo community and expanded my reputation as a junior officer who did not mind taking on additional responsibilities.

I had about six months left on my D.C. tour when the word went out to the OPNAV staff that the Presidential Inaugural Committee was looking for escort officers for the upcoming inauguration. I went to a meeting, and they outlined your responsibilities. The carrot was that you might be able to attend one of the Inaugural Balls depending on who you were escorting. There was some talk about Frank Sinatra and what an exciting time his escort had at the previous inauguration. We drew escort duty for an entertainment group that would be performing on the Ellipse, at a Youth Concert, and at one of the Inaugural Balls. Our job was to be their security escorts. You could not put a security badge on every performer, so we were their badges. We met them at their hotel, put them on the busses, got them admitted to the venue, waited in the wings for the performance to be over, then escorted them back to the busses and to their next scheduled event.

I had my first up close and personal Secret Service experience while serving as an escort. The group was performing for the President on the Ellipse. They were about to wrap up their performance when we realized that we did not have an escort on the other side of the stage. We needed someone there to make sure the group exited to the correct side of the stage by pointing them the other way if they were confused. It looked like we would have the group on the wrong side of the stage with no escort officer. I was maneuvering my way to the other side of the stage when I entered a tunnel that led to the stage. About halfway up, I realized that the tunnel also led to the Presidential Box. While rounding the corner, I came upon a very athletic-looking individual with an earpiece who was beginning to turn quickly toward me. The Secret Service Agent could see the uniform and the security badge and therefore did not put two rounds into my chest. Stopping, I pointed towards my security badge and then the stage. He nodded. Now standing in the wings, I was prepared to give the entertainment group the stop sign if they tried to come off the stage on my side. Everything went as planned. Our escort duties were cut short when the Inauguration Parade was canceled due to the cold.

It was about time to start negotiating for my next tour. I would be in the window to go back to a squadron, become an OINC, and do my Department Head Tour. Coming from D.C. I had to move somewhere so conceptually; I could go to a LAMPS squadron in Norfolk, VA, Mayport, FL, San Diego, CA, or Barbers Point, HI. The smart thing to do career-wise would be to go back to Norfolk and either HSL-32 or HSL-34. I was a known quantity in Norfolk, but I did not know anybody on the West Coast. I was leaning towards going to the West Coast. I had done both the Mediterranean and Persian Gulf cruises, and there was something about going to the Western Pacific (WESTPAC). I had always wanted to go to the Philippines, and there was the prospect of Australia, Singapore, Hong Kong, Pakistan, etc. The question became, how far west should I go, San Diego or Hawaii?

I plugged in with my former OINC, Bill Gibson, to see what he thought. He had been a junior officer detailer. He said, "If you are going to go west, go all the way to Hawaii. If you are not going to Norfolk, you are better off going where there are a lot of turnovers in the squadron, and nobody could homestead."

Homesteaders were those folks in Norfolk and San Diego who rotated from a squadron to the FRS, the LAMPS helicopter wing, AIRLANT, or AIRPAC, and then back to a squadron. They stayed in one location the whole time. If you were a LAMPS pilot in Hawaii, there was no follow-on aviation tour. That left Hawaii or

Mayport, Florida, and like I said, I wanted to go to WESTPAC. CDR Jim Crawford, my predecessor as NHA President and the OPS officer when I was in HSL-34, had recently been selected for aviation command and would be going to HSL-37 in Barbers Point, Hawaii, as the Executive Officer. I certainly did not want to go to Hawaii if he did not want me in his squadron. I talked with him, and he said it would be nice to have someone in the squadron he had previously served with. There were no promises of becoming an OINC or getting a Major (Operations or Maintenance) Department Head job. I would have to earn my keep if I decided to go there. I called the detailer to see if there were any slots in HSL-37 for me to fill in the period that I would be rotating, and he said there were. About a month later, I discovered that Jim would not be going to Hawaii. The reasons were unclear, but now I would not know anyone in the squadron at the CO/XO level.

I was unsure whether to tell my detailer that I was thinking about going to Norfolk rather than Hawaii, so I called Sandy Clark. Sandy and I had been in the FRS together. He was one of the sharpest guys in the LAMPS community and was very savvy about how the Navy worked. Sandy was now the helicopter placement officer at BUPERS in D.C. As the placement officer, he was the squadron's representative in getting officers assigned. He would work with the detailers, who were the individual officers' representatives. Between the two, they would try to balance the needs of a squadron against those of the individual who was looking for a squadron assignment.

Sandy knew who was in a squadron, who was leaving, who was ordered in, and most importantly, had access to their respective records and could judge how competitive they were for aviation command. I dropped by his office and explained my situation to him. He listened to my concern that I did not know the CO/XO or any of the department heads in HSL-37 and had no idea how strong the competition would be for a Department Head job, especially for a guy coming from the East Coast. To be competitive for Aviation Command at this point in my career, you had to have a successful OINC cruise and then do well as either the Maintenance Officer or Operations Officer. In some squadrons, the CO/XO would stack their squadron department head billets with their favorites and then slide them into OPS or MO. If you were not one of those favorites, you were a dead man walking. Sandy spent a good 15-20 minutes looking at the LCDRs assigned to HSL-37 and those who already had orders to the squadron. He looked at me and said, "you can be competitive there." I was not looking for any guarantees. He said my record stacked up against any of the LCDRs and that if I performed the way he knew I could, then I would have a good shot at either MO or OPS. I stopped by my detailer's office and confirmed that I wanted to be assigned to HSL-37.

CHAPTER VIII

GO WEST YOUNG MAN

HELANTISUBRON LIGHT THREE ONE (HSL-31)

THE ARCH ANGELS

FEBRUARY 1985 – JULY 1985

Going to HSL-37 was bound to be a wholly unique experience as I would be leaving the family in D.C. and going to San Diego to go through the FRS in HSL-31. I would be staying with Art Gallo. I had been Art's sponsor in Bermuda, and now he would show me the ropes in San Diego. He had also done a tour at HSL-37 when he left Bermuda and could bring me up to speed on what life in Hawaii would be like. My job was to eat, sleep, work, and get out of the FRS and to Hawaii as soon as possible so that the family could be back together. Trying to pay for two households, one in D.C. and the other in San Diego, would stretch our finances. I would also be in a much different role in the FRS. I was an experienced fleet aviator with three cruises, including one as an OINC. I had survived in the highly competitive world of Washington, D.C., and earned an M.A. from Georgetown by going to school at night. But I was an East Coast guy, and nobody knew me personally. Every day I felt like I needed to prove myself as an officer and an aviator. There was also the whole LAMPS East Coast vs. West Coast way of doing things. The East Coast was considered as being too uptight, anxious, and caught up in the rules and regulations, while the West Coast was more laid back and more inclined to make things up as they went along. My first introduction to these differences came right away in how you complete the items on the checklist in the aircraft.

During the checklist in the H-2, you come to a point where you must perform the engine checks. This involves moving switches and increasing/decreasing power on the engines. If one of the engines were to overspeed, the pilot at the controls would have to increase collective to load the rotor system and prevent the engine from overspeeding even further. On the East Coast, when you came to that point in the checklist, the aircraft commander at the controls would monitor what the copilot was doing. On the West Coast, the aircraft commander would transfer control of the aircraft to the copilot, then perform the engine checks himself. When the check was completed, they would transfer control of the aircraft back from the copilot to the aircraft commander. Later, in Hawaii, when I would be flying with someone who had been on the East Coast, we would specify in the brief which procedure we would do. Also, on the East Coast, you could do cross-deck approaches and landings to the ship. The pilot farthest away from the hangar could be shooting the approach. On the West Coast, the pilot closest to the hangar had to shoot the approach.

On the East Coast, we had an Integrated Weapons System Review (IWSR) pronounced Eye Wash, where the detachment would go to a different hangar and conduct operations and maintenance as if they were deployed. The primary purpose of the IWSR was to train the maintenance folks on how to do different major maintenance actions on the aircraft while the Technical Representatives were there to provide information and recommendations. The Tech Reps also knew all the shortcuts and what you needed to watch out for when performing maintenance on the aircraft. The detachment would practice removing and replacing an engine, a main rotor blade, a tail rotor blade, or how to replace a main gearbox or an intermediate gearbox. I hated the concept when

HSL-34 Det 1 had to go through it. It made no sense to me to take a perfectly good airplane apart and put it back together for practice. Inevitably these maintenance actions would require a functional test flight after the repairs were complete. You also ran the risk of breaking something and then being down while awaiting the parts. The General Electric Technical Representative and I had several heated discussions about how to tune and top the engines and the preventative maintenance to perform on the engines while at sea. After a while, I came around to his way of thinking. At the end of the IWSR, I had become a convert. The training proved invaluable on both my dets in HSL-34. The idea that you did not want to complete a complex maintenance procedure for the first time at sea made a lot of sense. So much so that I would later introduce a similar system in HSL-37 when I was the Maintenance Officer.

I was the class leader or "daddy rabbit" for my FRS class. As opposed to a class of ten like I had been a part of when I went through the FRS on the East Coast the first time, this time there were only four students in the class, Ensign Patsy Von Bloem, Ensign Steve Trainor, and Ensign Jim Hunter. Looking back, I regret that I did not provide more guidance or leadership to them. My focus was on getting through the FRS as quickly as possible and reuniting with the family in Hawaii. We also did not have a team-building experience like I had when we went to Key West for the flight training portion of the syllabus. Patsy was my Weapons System Training partner and would be going to HSL-31 Detachment Bravo, the sister detachment to the one I had been in on the East Coast. She would be deploying on the USNS Chauvenet T-AGS-29. We talked about my experiences on the detachment and jobs I had in the squadron when not deployed. Her assignment, like mine to det ALPHA, could provide a back door into the LAMPS community if/when women were finally allowed to be assigned to the Fleet Squadrons. Looking back, I do not think I did a particularly decent job in preparing them for their introduction to the Fleet. The fact that I was a new guy from the East Coast also affected me in that I did not want to be the guy who was always saying that back in Norfolk, we do it this way. I just kept my head down and progressed through the training. Steve Trainor and Jim Hunter would become squadron Commanding Officers later in their careers.

Things were progressing well in the FRS, and it became apparent that I would be able to take a week's leave if I wanted to. Linda and I discussed it and decided to combine the leave with a house-hunting trip to Hawaii. Her Mom and Dad would take the kids, and she would fly out of D.C. commercially. I would catch a military transport flight from March Air Force Base in California. We met at the recently refurbished Moana Hotel on Waikiki Beach. We knew that we would not be living out on the economy in Hawaii, so we went to the Government Housing folks to get an idea of our eligibility and how long we would have to wait for a house to come open. We were able to get on the housing list and looked at Pearl Harbor and Barbers Point housing. We toured the units and had a chance to look at the schools that the girls would be attending. We decided that we would live in Ewa Beach at Iroquois Point housing, and the kids would go to the school there. We also took the time to get a feel for the geography of the island and really enjoyed our time at Waikiki. It was good to firm up our plans for transitioning to living in Oahu, and the time flew by. The only good thing about saying goodbye was that I knew the next time I saw Linda and the kids would be when we were on our way to Hawaii.

The last couple of weeks in the FRS flew by, and before I knew it, we were on our way to Hawaii and about to join The Easyriders.

CHAPTER IX

ANOTHER DAY IN PARADISE

HELANTISUBRON LIGHT THREE SEVEN (HSL-37)

THE EASYRIDERS

AUGUST 1985 – FEBRUARY 1988

My arrival in paradise was everything I had hoped it would be. When I was on the OPNAV staff, we talked about favorite duty stations one day in the office. All the usual suspects were named, and it appeared that San Diego would be named the best place to be stationed. One of the guys in the office was a surface warfare officer who rarely voiced an opinion, and as a result, we modified a tv banking commercial to "when CDR Svenson speaks, people listen!" When he was asked about whether San Diego was the best place in the Navy to live, he smiled and said, "people who say San Diego is the best place to live have never lived in Hawaii!"

I did not have a job in the squadron for the first few weeks. I was waiting to be an OINC, but the detachment had not yet been formed. I tried to help anywhere I could and spent some time in operations, getting a feel for detachment schedules, pilot's and aircrewmen's reputations, and the syllabus I would have to go through. I started flying regularly; another good reason to hang out in operations, they write the flight schedule. I also was taking the time to get a feel for how the squadron ran and the dynamics between the CO/XO, department heads, and the OINCs.

The Skipper, CDR Mike Putnam, had a reputation as a tough guy to work for. He knew what he wanted, told you his expectations, and held you accountable. If you screwed up, he would let you know directly. The Skipper had no problem being the "bad cop." The XO, CDR Bill Fetzer, provided the buffer and was the "good cop." He would do his best to run interference and keep you from getting on the wrong side of the Skipper. I laid low and was happy when I was eventually told to establish my detachment. I would form up Detachment Four, who would eventually become known as "The Bounty Hunters" and whose motto would be "We live for the kill!"

I immediately liked what I saw when the detachment formed. The MO, LT Jim "Jaime" Gilchrist, would be going on his second cruise and was very intense and focused on doing the best possible job. A lot like I was but even more dedicated. He was always looking to improve himself personally and professionally. Because he was always on the move, the det sometimes referred to him as "the Shark" because as everyone knows, if a shark stops moving, it dies! OPS, LT Bill "The Beave" Thomas, was a lot more laid back and was the guy to provide comic relief no matter how serious the situation. Both the MO and OPS were good sticks, and they kept me out of trouble a few times when I was not as focused on my flying as I should have been. ADMIN, LTJG Gary "Body" Williams, was going on his first cruise; everything was new and exciting for him. It was fun watching him learn. The Detachment Chief Petty Officer was ADC Al Cornell, but I always referred to him as "El Jefe" or The Chief. He was an aviation machinist mate, so this would be the third time I would have someone from that rating as my crew leader. He was recently promoted to CPO, and it was interesting to see him grow into the rank. The MO and the Chief took a while to work out their relationship, but it was not long before they were in sync. The rest of the det were AT1 Miller, AE2 Neuwirth, AMH2 Spangenberger, AW2 Eaton, AD2 Seguin, AX2Morgan, AW3

Murr, AEAN Davis, ADAN Hobbs, and AMSAN Janet. Our aircrewmen were AW2 Eaton and AW3 Murr.

It would have been a dream to be assigned to one of the ships in Pearl Harbor, but I never did things the easy way. We would be assigned to a Fast Frigate out of San Diego, FF-1069 USS BAGLEY, as a part of the ENTERPRISE Battle Group. The Chief and I made a trip to San Diego to check out the ship and the aviation facilities. The CO of the ship was a surface nuclear-trained officer. As a nuke, he indicated that he wanted to be involved in all aspects of our operations and maintenance. I was concerned about this but felt that over time we would prove our professionalism and ability to do the job right and would be given more latitude. I was wrong.

We embarked on an exercise in the Hawaiian Operations Area along with the rest of the ENTERPRISE Battle Group. Following the exercise, the Battle Group would be departing for our WESTPAC cruise. One day our mission was to do ASW, and a U.S. submarine had been provided for the exercise. The exercise would take place on the Pacific Missile Range Facility (PMRF) off Barking Sands on Kauai. The sub was supposed to force contact, and then the exercise would begin. We had been assigned a search sector and were operating in it when the AW said he had pop-up radar contact. He gave me a vector to the contact, and we were soon overhead. We could see the periscope wake, and as we approached closer, we could see the submarine in the clear waters of Kauai. We marked on top and began a mad hunting circle so that we could develop course and speed for a simulated weapons drop. We reported the contact to the ship so they could pass it on to the Battle Group ASW Coordinator. The ship wanted to know how we knew that we were tracking the submarine. We said we have a visual. The ship asked, a visual on what? We replied, the periscope and its wake. The ship then asked, are you sure it is a submarine? We replied that we could see the whole damn submarine under the water. There was silence from the ship. We knew the contact had been reported to the Battle Group when the S-3s showed up, followed by the H-3s. Shortly after the gaggle arrived, we called it quits for the day.

The exercise supplied a good opportunity to work the det and the ship up simultaneously. It did not take long for me to conclude that the BAGLEY was nothing like the CONNOLE. The CO was an extreme micro-manager, and the XO was unwilling or able to provide a buffer between the CO and the Department Heads. The officers in the wardroom were good, and I had a good relationship with all of them. The Chief Engineer (CHENG) was constantly under pressure from the CO on the ship's maintenance. The XO had previously been enlisted and was nuclear trained. I had never seen such micro-management before and was about to experience it first-hand.

About halfway to the Philippines, we began getting main gearbox chip lights. The chips were more like fuzz initially, and we drained the gearbox, flushed it, did a penalty turn, and then began flying again. The frequency and composition of the chips increased over the next few days, so we had to down the aircraft. I was faced with a decision on whether to ask for a waiver for a one-time flight into Cubi Point Air Station in the Philippines or have the aircraft craned off when we arrived. One of the unwritten rules in naval aviation is that you never ask for a one-time waiver for a fly-off because the squadron would certainly deny it, and you would put the squadron into an impossible situation. You either flew it off, did not tell anyone, and accepted the possible consequences, or went by the Naval Aviation Maintenance Procedures. I decided to do the right thing, and we were craned off.

Main gearboxes were not easy to come by, and there had recently been a problem with them in the H-2 community. In fact, a sister det from the squadron deployed on a ship out of Japan was also down for a main gearbox, although theirs had just arrived in Cubi. I certainly did not want to ask for their gearbox because there was no estimated arrival time on the one, we ordered. The squadron decided that being with the Battle Group had a higher priority, and we received our sister det's main gearbox. I was surprised on the second day in port when the CO of the ship showed up at the hangar at the Air Station. I went over the maintenance manuals with him and explained the process of replacing the main gearbox. It was one of the more challenging maintenance actions that a detachment could be called upon to do. You took the top off the aircraft, including the hub, blade retentions, and the four main rotor blades. In our case, because we were shore based, we would take them off simultaneously as a unit rather than having to do each maintenance action separately. I gave the CO an estimated time of completion, and he said he would come to see us daily, which he did.

We finally put the aircraft back together and were ready for the functional maintenance check flight. I told the MO it was his aircraft, and he would do the check flight. We had to make several adjustments to the flaps on the main rotor blades for them to fly together and give us the proper RPM for autorotation. After several adjustments, we were close but still not where we needed to be. We were running out of daylight and needed to get the final adjustments done to beat the sunset. You are not allowed to perform a maintenance check flight after sunset. We were about two adjustments away from being complete and had about ½ hour before sunset. I told Jim and El Jefe to make a bigger adjustment and that I would do the flight check.

As I suspected, the time to complete the maintenance took a little longer than expected, and we ran out of daylight. I told the MO and the CPO I wanted to do the flight check after dark. If anyone did not want to participate in the flight, then I understood. The subsequent flight was perfect, and we nailed the blade track and autorotation rpm. I talked with Jim about it later and said I did not want to put myself in the position of ordering him to do the flight check after dark and that if anything went wrong, it was all on me, so I might as well be at the controls. He was still upset, but he understood my rationale for what I did. It was a little different doing a night maintenance check flight autorotation. I just added it to my bag of experiences.

I received an invaluable lesson in leadership from El Jefe while we were in Cubi. When we were planning the main gearbox maintenance, I told the Chief and the det that when the aircraft was up, the boys would have the rest of the time off while we were in Cubi. They put in incredibly long working hours and would secure late in the evening and adjourn off base for a little recreation "PI" style. A few hours later, they would be back on the airplane, working like crazy. One evening the det officers and I were also taking in the local culture. After a few hours, Jim and I decided we had enough for one night and reminded the two other pilots that we were supposed to attend an air operation brief first thing the next morning. We heard them come into our room in the Batchelor Officer's Quarters and daybreak was not far behind. When we showed up at the briefing, the OINC of the HSL-35 Detachment in the battle group got up, grabbed two trash cans, and gave them to my two junior pilots. "It looks like you are going to need these soon!" he said. Following the briefing, I told the two to return to the "Q," get a couple of hours of sleep and then report to the hangar.

When we arrived at the hangar, the Chief asked me where the ADMIN officer was. I asked him, what do you need? He said the ADMIN Officer was supposed to take care of some administrative stuff for a couple of members of the det. I said, ADMIN will be here in a few hours, and we will take care of it then. The CPO bristled and asked, "where is the ADMIN officer?" I tried to put him off, but he was insistent. Finally, I said, he was back in the "Q," and a little hung over. The Chief started in on me.

"All these guys," he said, pointing to our folks crawling all over the airplane, "were out last night too. But they came to work this morning and will be here as long as it takes. If an enlisted man can do it, an officer ought to be able to do it."

There was nothing I could say. I stared at my toes for a while and then told the Chief that I would personally get right on those admin things for the members of the det.

LAMPS Det Cubi had a great reputation on the East Coast for being able to help detachments with all their requirements when they pulled in. Having dealt with Lamps Mobile Support Detachment (LMSD) Sigonella in the Mediterranean, I had certain expectations about LAMPS Det Cubi. When we arrived, I had the usual laundry list of things that we needed to get done. We were given a base phone directory and pointed to a desk with a phone. When we asked who we should contact at these organizations, there was silence. No offers of facilitation or transportation. Things got a little better over the next couple of days, but nothing like Sigonella. I did not know it then, but I would have an opportunity to compare LAMPS Det Cubi with LMSD in Sigonella by the end of the cruise.

We left the PI and headed West. The Battle Group was under Emissions Control (EMCON) Alpha which meant we were not allowed to talk on the radio or use our electronic navigation equipment. BAGLEY was out ahead of the Battle Group, and our mission was to get out 50-75 miles ahead of the BAGLEY to ensure there was no surface traffic. We then would travel in an arc around the ship's proposed track and then turn inbound to intercept the ship. We would do this using dead reckoning, basing the intercept point on the ship's course and speed, combined with our course, speed, and wind effect. The timing would be tight, and I estimated we would intercept the ship with about 30 minutes of fuel remaining. Everything went well until we arrived at the estimated rendezvous point. There was no ship.

Every Navy pilot has experienced this feeling at some point in their career. The ship is never where they say they will be. You are supposed to compensate for that by giving yourself extra fuel, which equals extra time to find out where they are. We started an expanded square search pattern but did not have a visual on the ship. Options at this point were: break radio silence and ask the ship for a short count on the radio so that we could home in on them; ask the ship to bring up the Tactical Navigation System (TACAN), or have the AW use our surface search radar to try and locate the ship. I called the ship, but they would not answer. I informed them we were low on fuel and asked them to activate the TACAN momentarily. Again, nothing happened. At this point, we had about 15 minutes of fuel left. I told the AW to bring up his radar and take the minimum number of sweeps to produce a surface picture. When he did, the ship immediately came into the air and reminded us not to transmit. The AW said he had three contacts but could not be sure which one was the BAGLEY. I said make your best guess, and we are either going to land aboard the contact or ditch beside it.

Aircrewmen are taught that pilots will do dumb things and it is their responsibility to always know where the ship is. If the aircraft goes for a swim, everyone in the aircraft is also going to go for a swim. We briefed ditching and began to prepare the aircraft in case we could not land on the contact. BAGLEY came into sight, and we landed without further incident. That is if you do not count the ass-chewing I got from the CO, who had had his ass chewed by the Battle Group Commander. But I was standing there with my aircrew and aircraft intact and not in Davey Jones' locker, so it did not bother me too much.

A couple of days later, we lost an F-14 during EMCON ALPHA. Their navigation system had malfunctioned, and they ran out of gas. They were nowhere near where everyone thought they would be. The search lasted for the better part of a day. We were flying around the clock looking for them but did not hear anything on their emergency radio frequency. An A-7 picked up a weak emergency beacon, and an H-3 was dispatched to one of the nuclear cruisers in the Battle Group. The cruiser closed the estimated position at flank speed and launched the H-3 at maximum range. The crew was rescued, but the whole incident drove home how difficult it is to operate in EMCON ALPHA.

The ship pulled into Karachi, Pakistan and the BAGLEY was pier side. ENTERPRISE was anchored a few miles from shore. I was in the wardroom when the Shore Patrol Officer came on board looking for an aviator.

"What do you need?" I asked. He said that a sailor was extremely ill, and they wanted to get him out to the ENTERPRISE as soon as possible. He wanted cranials, goggles, and flotation gear for the sailor to use on the helicopter and said he would guide the helicopter into a landing spot near an area on the pier. We looked over the proposed landing site, and I said you are going to have to get all these merchants to take down their tents and move their goods out of the proposed landing area. He said the helo could land and would not have any problem with the merchants. I asked him two questions, are you a helo pilot, and what kind of aircraft would be making the pickup. He said he was an aviator, knew what he was doing, and that an H-46 was on its way in. I reemphasized the effect the H-46 would have when he tried to land, but the Shore Patrol Officer would not budge.

I said, OK, my guys will oversee the landing zone and direct the H-46 into land. We pushed the crowd out of the square and stood on the perimeter to keep people from wandering in. There was just enough space on the pier for the H-46 to land with the merchant's tents all around the periphery. We heard the H-46 coming, and I watched as my LSE guided him into the zone. When the H-46 flared to land, every tent around the square was blown over, and their goods scattered. I looked at the Shore Patrol Officer and smiled. We loaded the sailor aboard the aircraft, and after the H-46 took off, I told the Shore Patrol Officer that he was going to have to deal with the damage that had been done.

We began making metal in the new Main Gear Box, and it became obvious that we were going to have to change it out. The new gearbox would be coming into Diego Garcia, and BAGLEY would detach from the Battle Group to get it. The Battle Group Commander used to do a conference call every evening with the Ship's Commanding Officers. BAGLEY's CO used to have me sit in on them in case any aviation-related stuff came up. While we were en route to Diego Garcia, also known as "Dodge," the Battle Group Commander came up on the evening call and asked if any of the ships in the Battle Group had maps of the Mediterranean. We did not know why at the time, but it would be tied into the strikes that were going to occur in Libya in retaliation for the terrorist bombings in Germany. We were not allowed to send a message to the squadron to tell them what was going on. When we pulled into Diego Garcia, we adjourned to the Officer's Club for a little refreshment, but it was lunchtime, and they did not serve alcohol at lunch. But we had a satisfying meal. I went to find a phone to call home. Remember, I was not allowed to call the squadron and tell them where we were or where we were going. So, I called home and told my wife that I would be going where I had on my last cruise, and you remember what happened then, meaning CONNOLE, Mediterranean, and the Libyan shoot down. I asked her to call the Skipper at the squadron and tell him where we were going and that I would provide additional information as soon as I could. I did not tell the squadron where we were going and violate a direct order. My wife did.

We changed our second main gearbox while transiting to join the Battle Group. I had to repeat the process for the CO that we had gone through in the PI and supply updates throughout the day as we completed the work. It was like we had never done this before. I realized that this was how this surface nuclear trained officer worked. Seeing the aircraft in pieces all over the flight deck was a little disconcerting. Doing the maintenance without an overhead crane also made it challenging. We had to take the whole rotor head, hub, retentions, and blades off singly. Then we put the main gearbox in, followed by the hub, then each of the four retentions and four main rotor blades. At one point in the evolution, I briefed the CO that the flight deck would be clobbered while we did this maintenance and sent a message to the Battle Group that our flight deck would be closed. Sure enough, right in the middle of the maintenance, an H-3 flew by.

The next thing I hear is flight quarters, flight quarters. When I ran up to the bridge to see what was going on, I asked why we are sounding flight quarters. It will take me at least twenty minutes to get all the pieces off the flight deck, and the aircraft into the hangar, and I will lose at least four hours of work. I recommended that the ship tell the H-3 to come back later. The CO did not want to send the helo away because that would make us look bad. There was no response when I reminded him that we had told the carrier that we would be closed all day today for maintenance on our helo, and the H-3 should not have shown up in the first place. The carrier would not hesitate to send us away if it were not convenient for them. The CO would not change his mind. He told the ship's air controller to tell the H-3 we would be ready for him in 30 minutes. My frustration boiled over, so right there on the bridge, I said to the CO, "I guess it is true what they say about the BAGLEY."

“What is that?” He asked. My response was, “why plan when we can react?” The CO caught my tone and my disgust, looked at me, and said, “noted.” I left the bridge and went back to get the pieces off the flight deck and the helo into the hangar.

The test flight went without a hitch, and we proceeded to the Mediterranean. My wife told me later that the folks in Hawaii were surprised to see the BAGLEY transiting the Suez Canal along with the ENTERPRISE on CNN.

THE AUTHOR IN TRANSIT THROUGH THE SUEZ CANAL

I had gone to Hawaii so that I could do a WESTPAC and still wound up in the Med. The good news was that I was familiar with all the reporting requirements for LAMPS detachments, including our friends at LMSD. When we were en route to the Med, I told the other pilots on the detachment that the weather in the Med would be benign compared to what we had experienced in the Pacific. Upon arrival, the weather was the worst I had ever seen in two cruises to the Med. We operated for a few weeks before things quieted down to the point where we could have some liberty. Our first port of call was Monte Carlo while the ENTERPRISE pulled into Toulon. Monte Carlo was quite different from the Philippines, Karachi, and Singapore. I do not think some of the sailors had ever seen Caucasian women on the beach without their clothes on before. The beach bars did a booming business as the sailors enjoyed the view. Everything was more expensive, and we were on alert for any terrorist activities in retaliation for the Libyan air strikes.

We continued to have problems dealing with the ship. So, when the BAGLEY was scheduled to pull into port in Gaeta, Italy, I decided the det needed a road trip. We requested permission to shore base at Naval Support Activity Naples, Italy. This would put us about an hour and a half from the ship by road, which should provide us sufficient distance from the ship. The airplane would be at Capodichino (CAPO) Airfield, which the Italian Air Force, commercial civilian traffic, and the U.S. Navy used. The detachment would be staying in a civilian hotel near Agnano, about fifteen minutes from the airfield. The detachment quickly adjusted to the adventurous driving in Italy and were amazed when they learned that the mini bar in their hotel room was not included in the price of the room.

When the ship pulled into Gaeta, I went to the Sixth Fleet Flagship to see if I could get aviation maps and charts from the detachment that were normally aboard the ship. We had requested them when we were ordered to the Mediterranean, but they had not come in yet. Their det was in Naples, and there were no aviation publications on the ship. We should have driven down to Naples and gotten the charts and pubs, but I really wanted to get off the BAGLEY. I knew no one would approve a flight if we told them we did not have the correct charts. Looking at a regular map, we got an idea of what we should be looking for as we flew south, and we knew the tower at Capo would be up a standard Navy frequency. So, we took off, flew south, and heard various Italian Air Space

Agencies calling us on our Guard frequency. We explained who we were and where we wanted to go, and they cleared us through their air space. Eventually, we arrived safely at Capo.

We were greeted by the LMSD folks and the Kaman technical representative, who I had known since I came into the community. Jack King was regarded as "Mr. H-2" on the East Coast. The Chief was a little wary as he and the tech rep exercised their territorial prerogatives and manhood. The Chief said there was nothing wrong with the aircraft and we did not need any civilian Techrep telling us what to do. The Tech rep suggested they get a cappuccino. As they were drinking their coffee, Jack King engaged the Chief in casual conversation about the aircraft and things we had been doing maintenance during the cruise, especially the main gearbox. They agreed the main gearbox was a piece of trash and were soon talking like old friends. Jack showed the Chief a few tricks on H-2 maintenance, and we wound up spending more time working off gripes on the aircraft than we did flying. As I recall, the second day, the CO of the ship showed up, and Jack King took him aside and said what an excellent job the detachment had done in replacing the main gearbox while at sea. I hoped that would help with the CO's tendency to micromanage. It did not. I could not have imagined at the time that I would be the Commanding Officer of Naval Support Activity Naples eleven years later.

My worst moment as an aircraft commander came as we were in transit to the Philippines on our way home. The incident could have cost the lives of a couple of chaplains, found me willing to ignore the prohibition on transporting passengers at night, stretching how far I would fly with a main gearbox chip light and would result in my final showdown with the CO of the ship.

The mission was simple enough, transport a couple of chaplains to a ship. We would pick them up from the ENTERPRISE and take them to another ship. But there were a couple of mitigating factors. One, our old nemesis, the Main Gearbox, was making chips again. Two, it was getting late in the day, and it would be tight to get the chaplains to their ship before dark. How far did I want to go with the main gearbox, given our history with them over the course of the cruise? The range to the ship would be about fifty miles. What was our estimated time of arrival, less than fifteen minutes before sunset? I should have scrubbed the mission, but I elected to press on.

When we were about ten minutes from the ship where we were taking the chaplains, and the sun had already set, the ship called me to say that they were canceling the event. Angry, I should have known better and canceled the event myself long before. And as if to reinforce the point, the main gearbox chip light came on. The ship that was closest to us did not have a deck on which we could land. Now I was about forty miles from BAGLEY, in the dark with two unauthorized passengers, and wondering if the main gearbox was going to come apart. I declared an emergency and requested the BAGLEY close my position. The ship was slow to respond. Meanwhile, we prepared for ditching. A main gearbox chip light required you to land immediately. Because we did not have a place to land, we descended to about one hundred feet, slowed to fifty knots, and briefed ditching. If the gearbox made any noise at all or the pressure dropped to zero, I was going to put the aircraft in the water. I kept thinking how stupid it was that I had put myself in this situation. I began talking to the airplane in my head and said, do you want to swim with the fish? Get us back to BAGLEY, and you will spend this evening in the nice warm hangar. Then we will replace that nasty gearbox.

It seemed like it took forever to get to the ship, and I was getting increasingly agitated at myself for all the stupid things I had done on this flight. BAGLEY had not been particularly helpful and supplied a focal point for my anger at myself. Once I had done the paperwork, I went down to the wardroom, where the CO was having dinner with the Chaplains. I walked in and told the Captain I needed to talk with him right away. He asked me if it could wait until after dinner, and I said it was best if we talked right now. I was out of line, which was a serious breach of decorum in the wardroom, but I did not care anymore. We went to his cabin, and I told him I had violated several regulations during the mission because I wanted to make the ship and the detachment look good. I said I would not do that anymore. I felt that the ship had not supported us the way they should have during the deployment and that we were being micromanaged when we had proven that we knew what we were doing. It got heated, and eventually, the CO threw me out of his cabin. I figured I would be on my way home soon, but I did not care at that point.

The CO had previously fired several of his department heads, and the joke on the ship was that Surface Force Pacific had a desk reserved for former BAGLEY department heads. I had played a small role in the relief of the Chief Engineer previously on the cruise. The CHENG was super intelligent, and about the most dedicated surface warfare officer I had ever served with. He took everything seriously. He had to, with a couple of nuclear-trained micro-managers above him. One day we were talking, and he expressed his frustration with the CO's micromanaging. I asked, has he ever thrown you out of his cabin? The CHENG said no. I said he does not realize how much he is frustrating you, and every department head should get thrown out of the Captain's cabin at some point in their tour. He laughed and said he was not that kind of officer and did not think that would work with this Captain.

A couple of days later, I was in my usual position, in the rack, when Jim Gilchrist came in and said we are flying today. No, we are not, I said. We are not scheduled to fly today. Jim said we have to take the CHENG over to the oiler. He hit the CO on the bridge last night, and they are transferring him immediately. I bolted to the CHENG's room. He looked at me and smiled. I know what you are thinking, he said. This had nothing to do with our conversation the other day.

He said he had gone to the Captain with a plan of action and milestones to ensure that sailors did not spend too much time in their extremely hot engineering spaces. The engineers, snipes in Navy terms, are God's people who ensure the ship runs, the lights stay on, and there is water to shower in most of the time. Anyway, the CO had ordered him to produce this Heat Exposure Plan. The CHENG sat down with his Leading Petty Officers, then his Chiefs, and finally with the Division Officers to hash out a plan. It was complex, involved, and was a house of cards where if you pulled one out or changed the assumptions the plan was based on, the whole thing would come crashing down. The CHENG said he went to the Captain, who immediately began nitpicking on all aspects of the plan. The CHENG said, "at that point, I just lost it and slapped the CO in the face. It had nothing to do with our previous conversation." The LTJG who took over after the CHENG left did not seem to get as much guidance and direction as his predecessor did.

When the ship pulled into the Philippines, we went through craning off again. Fortunately, a main gearbox was waiting for us, and we had a lot of experience, so the removal and replacement went without a hitch. We were in the "Q," slang for BOQ, or Batchelor Officer's Quarters, and spent most of the day at the hangar, then went for a run, followed by a dinner and some refreshments out in Magsaysay or Olongapo. Jim Gilchrist had been trying to get me to have a massage at the BOQ for several days, and I had blown him off. I had heard about massages in the Philippines and was not interested. He said, no, this is a real massage. The day before we pulled out, I agreed to get a massage. Jim said, ask for

Nasty Nel. I did and was greeted by a Filipino lady who said to strip and get in the hot bath. She left the room and said she would tell me when it was time to get out. She called into the room and said to put on my towel and get on the massage table. Then she gave me the most incredible legitimate massage I ever had. She cracked my knuckles, ears, toes, and many other things I did not know that you could crack. Just when I thought this could not get any better, she walked on my back. I thought I was in heaven. When she finished, I walked into the Cubi Point Bar, where Jim handed me a cold San Miguel beer. Then I really was in heaven.

The night before we were going to pull into Pearl Harbor, the CO of the BAGLEY gave me my fitness report. While he thought it was good, I knew it was not. The only thing I could hope for is the Skipper back in the squadron would not forward it to the Bureau of Naval Personnel. We flew off the next morning and arrived at Barbers Point on the weekend. The Skipper, XO, MO, and a few other folks were there to greet us as well as the family. I knew the Skipper would go to Pearl Harbor to welcome home the rest of the detachment and that he would be talking with the CO of BAGLEY. I told him there had been issues on the cruise with the CO and that I had not handled them as well as I should have, and he could expect some negative comments about the detachment. When we got to the ship, the Skipper went to talk with the CO while I grabbed my personal things and assisted with the detachment offload. I never heard a word from the Skipper about the conversation with the CO of the ship, and the concurrent FITREP never made it into my service record.

When I got back to the squadron after some time with the family, I was burnt out. CDR Fetzer said that I would be taking over as the Training Officer while he figured out who he wanted to be his Maintenance and Operations Officers when he took over command. No promises. Do your job, and we will see what happens. I reviewed the existing training program, talked with the officers currently assigned to the department, and we produced some proposals to try to improve the training. We also revitalized the pilot's lesson plans on NATOPS and tactics. A couple of months went by, and the XO called and said he wanted to see me.

I entered CDR Fetzer's office, and I thought the conversation was going to be about how we could further improve training for the squadron. Once I sat down, he said he had been looking at the department head assignments and asked whether I had a preference. I said I had a feel for operations and felt that I would make a great operations officer. The XO agreed and said, you also have a lot of maintenance experience. I said yes, but I was not an engineer or a mathematics guy and was more comfortable training aircrew and producing the flight schedule. I gave him my complex analysis of how the H-2 operates, "Gas goes in, smoke comes out, and the rotors go round and round!" I finally decided to come clean. I said XO, the MO has a lot of responsibility. He owns about everything in the squadron, including airplanes, people, and supply. At department head meetings, it is always the MO who does the most dancing and is under the gun, and I am not sure if I could do that. CDR Fetzer smiled and said, I know you can do it. The light bulb finally came on. I was going to be the MO, no matter how much I argued about going to OPS. Doing the job as the MO would eventually lead to my screening for aviation command.

After the change of command, I took over as MO, and it did not take long to figure out that OPS and I came from different planets. OPS looked at the number of aircraft on the hangar deck and scheduled them all, regardless of whether they were ready to fly. His philosophy was that the MO's job was to maintain the aircraft, and OPS's job was to fly the crap out of the aircraft. This led to several "serious discussions" at Department Head meetings before CDR Dick Brown, the new XO, told me to turn it down a notch.

Eventually, OPS rotated out of the squadron and was replaced. CDR Mike Carnes was the new operations officer. Mike figured me out quickly. He would have his scheduler talk with the Chief in Maintenance Control every day before he roughed out the flight schedule. We went from maintenance *should be able* to provide X number of aircraft to maintenance saying they *will have* X number of aircraft. The flight schedule was much more realistic, and the folks in maintenance worked hard to provide the aircraft that we said we would be able to provide. Mike and I got along great personally, and that reflected in how smoothly our departments and the squadron seemed to run. We were able to sit down and talk things out.

One day the Skipper called down and said he wanted to talk to me in his office. As I went to see him, I reviewed in my mind the current state of the Maintenance Department and tried to guess what this meeting was about. I could have never guessed in a million years.

The Skipper told me we were going to modify four aircraft for a special mission, which he could not tell me about. He said we would be forming a multi-plane detachment onboard USS LEFTWICH (DD-984) out of Pearl Harbor. The other two aircraft belonged to detachments that would accompany the USS RANGER (CV-61) on ships based at Pearl Harbor. He also said that technical representatives from Kaman Aerospace Corporation would be arriving soon and would work with members of the three detachments to make modifications to the aircraft. We reviewed the aircraft assigned to the detachments to ensure they were the ones we wanted. They were the later models of the H-2, including Easyrider 55, the aircraft with which I had deployed.

The Kaman Techreps arrived in the squadron on the sixth of July, and we began to tear the aircraft apart. We had been told that we were going to incorporate Air Frames Change 326 (AFC-326). AFC-326 consisted of incorporating several weapons systems, increasing the electrical capacity on the airframe by installing a new generator, and extensive rewiring. We would install the AAQ-16 Forward Looking Infrared (FLIR) System, the ALQ-144 Infrared Countermeasures System, the ALE-39 Flare and Chaff System, and the hardware to mount M-60 machine guns. While the maintainers were busy on the aircraft, the pilots and aircrew were learning how to operate these systems in a hostile environment. The two-primary aircraft for the mission were scheduled to deploy on the LEFTWICH, but the aircraft assigned as backups with the RANGER Battlegroup would be leaving first.

While modifying the aircraft, I began to get calls from the maintenance folks on the staff at ASWWINGSPAC in San Diego. No one had told them about the modifications. I described what we were doing. They asked who authorized the changes. I said they had come from OPNAV and AIRPAC. It seemed like I would get a call from the wing every other day. It was like Groundhog Day. What are the changes, and who authorized them? I was waiting for the Wing Maintenance Officer to show up at the hangar, but he never did. At my level, I never heard from OPNAV, AIRPAC, or anyone else.

At one point, parts of the aircraft were scattered all over the hangar deck, and I had some real doubts about how this was all going to work out. I had images of the new generator melting, the wiring shorting, or the newly installed gear not working properly. Two weeks flew by, and the modifications were done. Now it was time to do the Functional Check Flights. We started the aircraft. The new generator assumed the load of the additional equipment, the wiring worked, and the equipment came online as designed. We completed the check flights with no major discrepancies. It was an extraordinarily successful evolution on the part of all involved people. The number of things that could have gone wrong was astronomical, but the dedication of the folks from Kaman and the Easyriders made it work. The SH-2F "Attack Wolf" was now ready to deploy.

The RANGER detachments had completed their training and were getting ready to deploy. LCDR Wayne Lachowicz had deployed with me on CONNOLE. LCDR Lew Wolfrom had been assigned to HSL-30 Det Alpha after I had left and had done hydrographic research work in the Caribbean. One was a Coyne trained man, and Lew had worked for me in training when I returned from deployment. Lew had also taken over my old det, the "Bounty Hunters," and would deploy with Chief Cornell and some of the folks with me in the Mediterranean. I felt close to them and a little anxious about where they were going and what they might be doing on deployment. Before they got in their airplanes, I walked up to Wayne and Lew and said, "a successful detachment is one where you bring the aircraft and all the people home." The people would come home, but the aircraft would stay in the Middle East.

Once they departed, we focused on the LEFTWICH detachment. The ship would be modified to make sure the two H-2s would fit. I had done a similar experiment with a KIDD Class Destroyer back in Norfolk while in HSL-34. There was not a lot of room to spare when the aircraft were in the hangar. Every OINC tells their det that the most dangerous thing you do on a LAMPS detachment daily is to move the aircraft in and out of the hangar. There is little margin for error, in some cases inches. Multiply that by two aircraft, with no additional space, and you have a challenge. Jay Butler was the OINC of the LEFWICH detachment, and he was always afraid that I would replace him before they left on the cruise. I had just completed an extraordinarily successful cruise, deployed to the Persian Gulf, been in the Med twice when we had dust ups with the Libyans, fired an M-60 machine gun from an H-2, and was the senior Lieutenant Commander in the squadron. Jay spent a lot of time in my office, and we tried to anticipate all the challenges he would face in a first-ever multi-plane LAMPS Mark I deployment. We tried to tailor the maintenance folks assigned to his detachment to ensure we had the right expertise and eliminate potential personnel problems. With the three detachments deploying simultaneously, we stripped the squadron of our most experienced maintenance folks to ensure the deploying detachments' success.

While all this was going on, no one in the LAMPS community seemed to be aware of it. But senior folks in the chain of command were paying attention. We got a call one day that the Commander in Chief Pacific Fleet (CINCPACFLT) was coming over and wanted to see one of the aircraft. The most difficult part of the evolution was getting the M-60 machine guns out of the station armory. Seeing the Skipper and the OINC brief the Admiral on the aircraft was great. The dets went out and did a great, professional job. They set the standard for the use of the Attack Wolf for the rest of the community. When their cruises were completed, they transferred the airplanes in the theater to detachments from San Diego.

At the time, the XO CDR Brown wrote a point paper that proposed that the AFC-326 birds be put into a newly established LAMPS squadron. He said if you did not want to do that, then pick one of the existing squadrons and give them all the birds. The equipment and supply requirements for these aircraft were unique. You were ordering parts directly from the manufacturer rather than through the Navy supply system. There was no training pipeline on how to use and maintain the equipment. It was all technical representatives, word of mouth, and on-the-job training. CDR Brown felt the best use of the assets would be to keep them in one place where the institutional knowledge could overcome the lack of an established Navy Program of Record. I thought the XO's point paper hit the nail on the head. We heard back that if you put all the Attack Wolves in one squadron, no one would want to go to any of the other LAMPS MKI squadrons. Not the kind of forward-thinking you would have expected from the LAMPS community. Over the next couple of years, the aircraft remained deployed and rotated among the San Diego squadrons. Because they remained deployed, problems with training, maintenance, and spare parts would continue, as I discovered a few years later.

I loved being the maintenance officer and enjoyed meeting the challenge of putting a superior product into the air daily while ensuring that the detachments we deployed were as well-trained as possible. I embraced the things that I had feared when CDR Fetzer and I talked about me becoming the MO. I still did not understand exactly why helicopters could fly, but I loved taking them out for check flights after they had been repaired. I did not mind the responsibility for the people, airplanes, supply, and meeting the challenge of the daily flight schedule, or that I was usually the target for the most pointed questions at every department head meeting.

On every detachment, the OINC and the MO were usually functional check pilots (FCP). I felt my job as squadron MO was to ensure they were professionally trained and prepared to accept the responsibilities of being a good FCP. I used to tell the first tour folks who were being considered as FCPs that the job wasn't for everyone. It meant taking a sick airplane out and torturing it to make sure that it had been adequately repaired. You were often right on the borderline of exceeding aircraft limitations to ensure the maintenance had been done properly. If you were not prepared to accept that discomfort, which was necessary to do the job right, then you should not be an FCP.

I especially loved the relationship between the maintenance Chief Petty Officers and the Maintenance Material Control Officer

(MMCO). The Chiefs knew that my Dad had been a CPO, and they gave me the benefit of the doubt. The MMCO (CWO4 Emile Rosario) was a very experienced Warrant Officer who was always a few steps ahead of me. He was very patient in educating me on what the MO needed to be worried about and what the MO needed to let him do, which was about everything. One of his idiosyncrasies was that the angrier he got, the softer his voice would become. When he was enraged, it would be barely above a whisper. On those occasions when I heard whispering outside my office, it was my job to rescue the poor SOB, who was the subject of the CWO's attention. I would try to get them away from the Warrant as quickly as possible and tell them to fix whatever he was talking to them about immediately. He always referred to me as the boss. Between the Warrant and the Chiefs, the squadron's maintenance efforts were on a solid track. One of my saddest days in the squadron was the day I was sitting in the MO's office with my relief, and the Warrant walked in, looked at the guy who was now sitting in my old chair, and said, "Boss, here is what I recommend we should do." I was crushed. I knew that the loyalty was to the institution, not the individual. But it still hurt like hell.

I wanted to go to Newport and the War College, but there was a period where there was a possibility of going to the Middle East on the staff of MIDEASTFOR to coordinate the support for the Attack Wolves. It took a few weeks for that situation to sort itself out and a more senior LAMPS officer went to the job. I would attend the College of Naval Command and Staff at the Naval War College in Newport, Rhode Island. I was going to be a "phased input," which meant I would start in March instead of in the fall. The family and I began trying to prepare ourselves for going from laid-back and warm Hawaii to the prospect of a very structured and cold New England.

CHAPTER X

IT'S ONLY A LOT OF READING IF YOU DO IT!

COLLEGE OF NAVAL COMMAND AND STAFF

NEWPORT, RHODE ISLAND

FEBRUARY 1988 – MARCH 1989

"It's only a lot of reading if you do it." That was my introduction to the College of Naval Command and Staff (CNCS) in Newport, Rhode Island, also referred to as the junior course. The other saying in the Navy was that you needed to be selected for the War College, but you did not necessarily need to attend it. Selection indicated that you were doing well in your Naval career. The CNCS was for O-4's from the services as opposed to the O-5 and O-6's who were in the senior course, at the Naval War College. I had liked Newport during the brief time we spent there before deploying on the USS CONNOLE. I also wanted the opportunity to compare the course of instruction in a military college to that I had acquired at Georgetown.

The junior class was much more laid back than the senior course, and my impression was that we had a lot more fun. Most of the folks in the senior course were post-command and biding their time until their next assignment. For them, it was an opportunity to decompress from the rigors of Command while waiting to see if they would be selected for Captain. The captains in the course were waiting to see if they would be selected for major Command.

My contemporaries in the junior class had come from or were awaiting assignment to Department Head tours or Command. I was one of the most senior members of the class and I was waiting to see if I would screen for Command of a LAMPS helicopter squadron. I quickly settled into the routine of classes, lectures and reading. One of the benefits of attending the War College was that you were exposed to Navy officers from across the warfare communities of surface, submarines, and aviation, as well as folks from the Supply Corps and the Judge Advocate General Corps. Then you threw in the other services with their corresponding specialties, folks from other Government Agencies, and you began to get a better feel for their respective contributions to the national defense.

The Air Force folks believed that they could settle any conflict by themselves. It was a matter of the proper application of air power, which would settle the conflict without the necessity of the Army or Navy getting involved. The Army was all about the coordination and synchronization required to win any great land conflict, principally in Europe. The Navy tended to view Asia as the most strategically difficult theater, with the greatest economic impact, and that if the Army and Air Force just got out of our way, we would be able to assure victory through dominance in the air, on the sea, and under the sea. We also had these Marine folks who loved to break things once we put them ashore. These viewpoints made for some very lively discussions over the beverage of your choice, but also helped us to understand the unique capabilities and cultures that the individual services brought to any operation.

Before I left HSL-37 the CO, CDR Brown, a Naval War College graduate, had already introduced me to a couple of concepts that I would learn about in Newport. One day as the MO I walked into his office and started outlining a problem I was having with the idea that he would provide guidance or a recommended way ahead. His response was brief and succinct, "I'm not putting that monkey on my back." I was taken aback, and he laughed and said you will learn all about that when you go to Newport. It is a case study on decision making and problem solving. He said, "now take your monkey and get out of my office." The lesson was to not accept someone else's problems but force them to solve the problem themselves. He also said that Newport would change the way you look at things when you do the Torpedo Factory Study. The case study called for you to recommend which manufacturer should make torpedoes for the Navy. You started by producing a list of what you needed to be able to do to manufacture the torpedoes. You identified the crucial factors necessary to the manufacturing process. You would assign a "weight" or relative significance to each of these factors from most important to least important. A number was assigned to each factor from 1 to 10 based upon relative importance. You added up the numbers and you would have a numerical solution as to which company was the best choice to manufacture the torpedoes. You also learned very quickly that you could "skew" the results depending on what you decided were the most important key factors, and to which you assigned the most weight. The case studies were always fascinating, but you were given all the facts and then had to make a decision/recommendation. To me, the hardest part when I was out in the fleet was always trying to determine what the facts were. Once you had determined all the facts, the decisions tended to be easy.

It was quite an adjustment at the War College, going from an operational or tactical focus to a strategic viewpoint. Your job was to learn. You did not have any other duties or responsibilities. It took about three months to beep down from squadron life. The courses were divided into Strategy and Policy, National Security Decision Making, and Joint Military Operations. My electives were Military Media Relations, Military Justice and Administrative Law for the Commander, and the Defense Planning System. I enjoyed the reading, and I did it all. We all took the Defense Planning System to fulfill a Joint Professional Military Education (JPME) requirement. I would have two degrees in National Security, one from Georgetown University, and the other from the Naval War College. The emphasis in the military was towards compliance with the Goldwater/Nichols Act, which called for officer's to be educated in joint operations. I would later be designated a JSO because of my education and previous operational experience on the OPNAV Staff. The JSO designation would play a critical role in my future assignments and duty stations, but I did not have a clue at the time as to its significance.

The lecture hall at the War College was referred to as "the big, blue, bedroom." The hall was big, the seats were blue, and there was a tendency to fall asleep when the lecture was not particularly interesting. It was funny to see how my fellow classmates would ask their questions of our speakers. There were those who wanted to impress everyone with their knowledge and intellect. Their questions tended to be exceedingly long and clearly emphasized how smart they were. There were the folks who made sure to ask every lecturer some kind of question to show how involved they were. I was once chastised by my classmates when I violated my personal rule to never ask a question of a guest lecturer in that public forum. The question was something like, "what do you think about X?" No framing the question, no trying to demonstrate how I knew all sides of the issue. Just a straightforward inquiry.

During the summer, the War College hosted Global War Game 88. I was an aide to the command group. This war game was dedicated to figuring out how you would cease or end the war. It was particularly challenging, and I was amazed how seriously the players, which included senior folks from the Congress, Office of the Secretary of Defense, and the military took their roles. Congressman Newt Gingrich of Georgia's role was as President of the United States. At one point in the game there was a conflict over a recommended course of action. The DOD and military role players were pushing extremely hard for a certain recommendation. "President Gingrich" indicated he did not want to go along with their recommendation. When the DOD/Services pushed back, "President Gingrich" replied that they should either execute his order or have their replacements report to him immediately. It was like all the oxygen was sucked out of the room. At that moment it did not feel like a game, and I had to tell myself that it was not real. Global War Game 88 was my first exposure to how the "great men" operated in the national security decision making world.

Not everything in Newport was work. The most magical moment in Newport occurred when we attended the Christmas Ball. The event was held in one of the mansions in Newport and was a formal occasion, which meant mess dress. Linda and I attended and had a wonderful time. When we left to go home, light snow was falling, and it was one of those movie moments where everything was simply perfect.

One of my squadron mates and a predecessor as Maintenance Officer in HSL-37, CDR Jeff Keho, was also at the War College and we had him over for dinner around Thanksgiving. Jeff had been selected for Command and was scheduled to go back to HSL-37 as XO in the spring. During dinner he told us he was talking with the detailers about not going to HSL-37, but to VXE-6, the folks who flew C-130's and UH-1N's in supporting the scientists in Antarctica from a base in Christchurch, New Zealand. He said the detailers were giving him a tough time, and he could not go from a LAMPS fleet squadron to a support squadron. He felt his arguments were sound and expected the detailer to eventually give in and order him to VXE-6. I had not been selected on my first look for aviation command, and the board would be meeting in the spring. I did not know if I would be selected, but I now knew that there was a hole in Hawaii that would need to be filled in the spring by one of the guys who was selected.

After graduating in March, I would be up for orders. I did not want to move the family until the summer and started talking with the detailers about where I would go next. I received verbal's to go back to OP-906 in the summer, pending the outcome of the aviation command screen board. The folks at the War College agreed to let me stay until the summer and I would do odd jobs for whatever department could use me.

My class had graduated in March, and I was providing administrative support at the War College wherever necessary while waiting for my kids to finish school in Newport. One day I was in the student lounge when one of my former instructors from HSL-30, and the Key West experience, CDR Frank McCarty walked up to me and said, "congratulations Skipper!" That was how I discovered I had been selected for aviation command. I could not believe it. I was elated at being selected, and disappointed that I could not share the experience with my War College classmates or any of my LAMPS contemporaries. I had received this great news. The accomplishment of a lifetime. There was no one to brag too about it. The family was happy to hear it, but the excitement of my being selected was soon overwhelmed by the question of where are we going to go? The family wanted me to find out where that was as soon as possible.

Fortunately, I had served with my current detailer, CDR Bill Young, when we were stationed in Bermuda. I called him and asked, what are my options? I had been an SH-2F Mark I guy but he said I could also be assigned to an SH-60B Mark III squadron. Because I would be leaving Newport, I could elect to go to San Diego, Norfolk, Mayport, Florida, or back to Barbers Point, Hawaii. As I considered which community I wanted to be in, for me it all came down to what my credibility would be as a Commanding Officer. I knew the SH-2F as well as anyone in the community. I had deployed on Knox Class Frigates and Spruance Class Destroyers. I had been successful as an OINC and as a department head. I could look any OINC or Department Head in SH-2F community in the eye and they would know that I had been there and done that. The CPOs and enlisted folks would also know I had credibility when I asked them to do things. That would not be the case if I went into the Mark III community. I would be the new guy on the block and would not have any operational experience with the aircraft while deployed. Obviously, a lot of previous Mark I folks had succeeded in the Mark III community. But it did not feel right to me. So, I decided to stay in the community. Next decision point, where do you want to go?

I called Bill Young back and said I wanted to stay in the H-2 community. I knew that there was a hole at HSL-37 in Hawaii that would need to be filled in May but did not know if someone had already been designated to fill it. Bill asked me if I wanted to go back to Hawaii. ASWWINGSPAC had set some requirements for the HSL-37 job. You had to be available immediately (the Change of Command was scheduled in May), current, or recently current, in the aircraft (there was no time to spare in getting through the FRS), and you had to have been stationed in Hawaii before. There were only two guys who could immediately fill the slot under those conditions, Art Gallo was currently the XO of the West Coast FRS, HSL-31, or me. Art, Bill Young, and I had all been stationed together in Bermuda.

My concern in returning to HSL-37 was that I had departed the Squadron the year before and wondered, is it too soon to return? There would be folks in the Squadron who had worked with or for me as the Maintenance Officer. I wondered if it would be a problem for them to accept me in my new role. While I had served under the current Skipper of HSL-37, CDR Dick Brown, I did not know the XO, CDR Scott Steele. I also did not know if CDR Steele had any preferences about who his XO should be. I called the Squadron and talked to CDR Brown about my misgivings. He assured me that it was not too soon to return to the Squadron and that CDR Steele would welcome me as his prospective XO. I asked CDR Brown to run my assignment by CDR Steele to see if was ok with him. If CDR Steele had someone else in mind, I would move to another squadron. We concluded the conversation, and CDR Brown said to call him the next day. When I called back, Skipper Brown said he and CDR Steele would welcome me back to the Easyriders. I called and confirmed the assignment with Bill Young. I went home that night and told the family. The good news was they would be going to Hawaii in June. The bad news was that I would be leaving immediately.

CHAPTER XI

JOURNEY BACK TO PARADISE

HSL-31 ARCHANGELS

MARCH 1989 – MAY 1989

Art Gallo welcomed me to HSL-31, and I checked in with Skipper Putnam, who I had served under in HSL-37. The syllabus was designed to get me requalified as quickly as possible and out to HSL-37 for the Change of Command in June. I had to go to an abbreviated CO safety course at the Naval Post Graduate School in Monterey, California, an abbreviated syllabus to requalify in the aircraft and update my water survival qualifications. Safety school was a breeze, and getting comfortable in the aircraft was no sweat, but the Water Survival Course presented me with a challenge.

Never the strongest swimmer, I was always able to "bludgeon" my way through the water survival course. My two greatest challenges were demonstrating proficiency in the different strokes and qualification in the 9D5 or helicopter dunker. You had to pass the swim test to get to the dunker, and if you failed any portion, you would have to come back the next week. I did not have a week to spare, so it was literally sink or swim, if necessary, to pass the course. As I recall, I did not prepare myself well for the event. The night before, I had consumed quite a bit of the beverage of my choice, probably with Art Gallo, and so I reported to the pool a little worse for wear. No problem, I thought, I can handle this.

For the swim requalification, you had to swim in full-flight gear and demonstrate your proficiency in the breaststroke, backstroke, and crawl. I was particularly good on my back, not so much with the breast or the crawl. I had completed the breast and crawl portions and was on my back in the final twenty-five yards when I managed to scoop a bunch of water into my mouth. I immediately began to choke and started to flounder. The instructor tapped me with the dreaded stick and ordered me out of the pool. I had not been paying attention to the brief and thought we would go to the dunker, and I would have an opportunity to come back and repeat the strokes. The instructor said, "no way Commander, if you don't pass the strokes, you don't get to do the dunker, and you will have to come back next week." That was not an option, so I went to the back of the line to wait for my next and final shot at passing the strokes. I jumped into the pool and went through the first two strokes again with no problem. Flipped over onto my back and at the same exact spot as before, I shoveled water into my mouth again. I knew there was no next time, so I decided to choke and hold my breath while pulling for the pool's end as hard as I could. As I was pulling with all my strength, I thought about having to tell CDR Putnam and Art that I had flunked the swim qualification and all the possible repercussions that would come from that. I was totally focused on pulling as hard as I could while thinking about how I would explain my failure. I was so focused that I could not see or hear a thing, and I was surprised when my helmet hit the wall at the end of the pool. I looked up and was overjoyed when the instructor gave me a thumbs-up. After crawling out of the pool, I stood there gasping for breath. The instructor looked at me and said, "Class, we will continue to the dunker as soon as the Commander has recovered his breath."

So now it was on to the Helo Dunker or 9D5. Think of a gigantic 55-gallon drum that you can strap a bunch of folks into, roll it upside down, and then sink it. Then the people inside must orient themselves, unstrap, and exit the dunker. You must do this several times, and the events are all graded. If you cannot do it successfully, then you do not get requalified, and you cannot fly.

Did I mention that you also must do this blindfolded? There are two positions in the front to simulate the pilot and copilot and several positions in the back to simulate passengers. Egress from the pilot or copilot stations is much less complicated than exiting from the back because you exit through the doorway right next to you. If you are in the back or cabin area, you must try to get out simultaneously with three or four of your closest friends. Best case is for your check ride to be from the front. Worst case is to be in the very back of the aircraft, where you have the farthest to go and must deal with the other blindfolded folks trying to get out. I do not know if it was intentional, but you can guess where my final check ride position would be, the very last seat in the back of the simulator.

By the time of the check ride, you have done this several times, so you are confident you can do it one more time. With our blindfolds in place, we gave a thumbs up and were suddenly dropped into the water. The dunker began to roll over, and I figured I had a second or two before going under, so I prepared to take a big breath. Unfortunately, I had miscalculated the roll rate and got about a ¼ of a breath before my mouth filled with water. It was just like the pool all over again. I said to myself, do not panic. You still can do this.

Once we were upside down and all the simulator motion had ceased, I oriented myself and released my lap belt. As I moved hand over hand down the fuselage, two of my contemporaries took turns kicking the crap out of me as they made their way out of the simulator. Once again, I told myself, you can manage this beating because you do not have an option. As I moved down the fuselage, I was probing with my hand to see where the exit was. I finally reached out, and there was no resistance. I swept my arm to find out where the doorway was when I hit the "funny bone" of my elbow against the door frame. One of my arms is now completely useless, and I must get to the surface on one arm, with a mouth full of water, no air, and having had the crap kicked out of me. Once again, the images of Skipper Putnam and Art flashed through my head. I pulled for the surface with one arm and all the strength I had left and was surprised when I broke the surface of the water. I looked up to the instructor and was ecstatic when I saw the "thumbs up" that indicated I had passed my final ride in the dunker. I returned to the "Q" that night feeling like I had gone ten rounds with Muhammed Ali.

I finished the flight portion of the FRS in three weeks. The highlight was flying to Palm Springs with Art on my instrument check. We had come a long way from being Ensigns together in Bermuda. We were going to be Skippers' of a LAMPS squadron, Art in San Diego and me in Hawaii.

CHAPTER XII

WHAT A DIFFERENCE A YEAR MAKES

HSL-37 EASYRIDERS

MAY 1989 – JUNE 1990

EXECUTIVE OFFICER

My first impression upon reporting back to HSL-37 was how much the squadron had changed over the time I had been in Newport. When I left the squadron, the majority of the department heads and senior lieutenants in the squadron were married and either had or were starting families. When I returned, most of the officers were single or recently married. The squadron had a different vibe about it. Not bad, simply different. My family would be finishing the school year in Newport and would move back to Hawaii in the summer. So, I got to experience Hawaii as a geographical bachelor for a few months. Living in the BOQ was different. The commute to work took two minutes, and I spent a lot of time in the squadron because there was not much else for me to do. There were still some holdovers from my previous tour, but all the department heads had changed.

Over the next two years, my leadership skills would be tested as the squadron worked through two accidents and went to war in the Persian Gulf.

The first thing I needed to sort out was the relationship with Skipper Steele. We had never served together and really did not have a feel for one another personally or professionally. I had always wondered about how much planning by the Bureau of Naval Personnel, or the Functional Wings, went into matching a CO and XO's personalities and leadership styles. I would learn over time that there was extraordinarily little. I had been in squadrons where the CO was the nice guy, the XO was the hammer, or the CO was a hard ass, and the XO was easygoing. Those personalities carried over when there was a change of command, and the XO would assume a role to complement the Skipper.

XO OF HSL-37

Skipper Steele was a West Coast guy and was very laid back when it came to his management style. It was tough to read him as to what he wanted to be done or how strongly he cared about a given issue. He was more inclined to go with the flow than to shake things up. I was more confrontational and oriented toward how the East Coast did things. Over the years, I realized that I was either committed to something and would not easily compromise, or I had no strong feelings and was amenable to the proposed solution. I had to rein in some of my natural instincts to be more aggressive and confrontational because that was not what the Skipper wanted, and it would only confuse folks in the squadron.

I had a great deal of respect for the way the Skipper Steele handled our aircraft accidents. I would complete the trifecta of being the subject of an accident mishap board, being on a mishap board, and convening a mishap board.

The first mishap could not have come at a more traumatic time in our personal life. I had arrived home from the squadron to see an ambulance outside our house. When I went in, Linda told me our youngest daughter Erin had run through a plate glass sliding door that led to the lanai, shattering the door and severely cutting one of her legs. Erin was sitting on the kitchen counter talking to the emergency folks when I walked in, and her leg was wrapped up, so I did not have an opportunity to see how bad it was. They loaded her in an ambulance, and I followed them to Tripler Army Hospital. When I walked into the emergency room, they uncovered her leg, and I almost passed out. The leg was cut down to the bone. They injected Novocain around the wound, and as I watched, I became very lightheaded and realized I had to sit down before I passed out and fell on the floor. We were in the emergency room for a couple of hours before they discharged her, and we returned home. The doctors felt her leg would be fine, but she would have a nasty scar.

These were the days before cell phones, but we did have an answering machine at home. There were several messages on the answering machine, and I assumed they were folks calling to check in on how Erin was doing following our visit to Tripler.

The first call was from the Squadron Duty Officer asking me to call the squadron.

The second call from the Duty Officer was more urgent and asked that I call the squadron as soon as possible.

The third call from the Duty Officer was simple and direct, "XO, call the friggin' squadron right now!"

Subconsciously I am sure I knew that there had been an aircraft accident, but I did not want to admit it to myself. I called the SDO, and he confirmed that we had lost an aircraft. I set the land speed record for travel between Iroquois Point to Barbers Point that night. I joined the Skipper in his office. The first report was that the aircraft had crashed at sea and that two of the crew had survived, the Aircraft Commander and the Aircrewman. The ship was conducting a search to find the junior pilot, an Ensign who we believed did not make it out of the aircraft.

Our initial efforts were to report the accident, form a mishap board, and get them started on their journey. The crash had occurred off Okinawa. A helicopter squadron on Guam would provide the senior member of the board. We would send our Safety Officer and a couple of other pilots as members of the Accident Mishap Board.

Meanwhile, Skipper Steele would notify the widow and her baby son. I do not think that any CO is ever prepared to make the notification to the next of kin. I felt guilty because I was relieved that I would not have to do it. It was getting late, and we discussed whether to wait until the next morning to make the notification. The Skipper was worried that the widow would find out unofficially, and so he made the decision to go that evening. I remained at the squadron and went through the accident checklist to make sure that we had notified everyone in the chain of command and that we were following the proper protocols and procedures. Skipper Steele arrived at the Ensign's quarters, and there was one of the squadron wives walking up to the front door. The Skipper intercepted her and told her to go home. I know the Skipper, and I would never have been able to live with ourselves if we had waited until the next morning to make the notification and found out the widow already knew. It has always amazed me how fast the information about an accident makes its way through the squadron, even though no one is supposed to talk to anyone about it.

The next day we learned that the Ensign's family would be coming to Hawaii. We started to plan for a memorial service and tried to determine how we could assist and support the family. Linda was trying to care for our daughter, who had just gone through a traumatic experience, while simultaneously providing comfort and assistance to Ensign Ralph Barile's family. She arranged to take the Barile family on a tour of the island in our van. She would load Erin into the van and then the Barile family. Julie Barile, the new widow, was genuinely concerned about Erin's leg and how she was managing the trauma associated with going through a glass sliding door. She told Erin that if she had a scar on her leg, she should tell everyone that she had been bitten by a shark while swimming in Hawaii. We marveled at Julie's ability to help someone else deal with trauma when she had so much going on in her own life.

The role of the Skipper and XO's wives is not really defined in Navy Regulations, especially when it applies to an aircraft accident. Because Skipper Steele was not married, Linda assumed the role of acting as a sounding board and going between the family and the squadron. She tried to gage what the family members wanted or needed and then tried to meet their desires. The members of the wives' club provided food and tried to make sure that someone was with the family as much as possible while respecting the family's wishes. The Squadron Casualty Assistance Control Officer, or CACO, was assigned to provide coordination between the family and the Navy. He assisted the family in any way they needed and explained their Navy benefits.

We began to get information from the Accident Mishap Board about the circumstances surrounding the accident. The ship was conducting small boat operations, and the aircraft was orbiting while waiting to land. The aircraft made several passes by the ship at low level and high speed. The ship CO asked the aircraft commander if that was the best he could do. The aircraft commander then executed a rotor over where he did a zoom climb, rolled the aircraft forty-five degrees, lowered the collective, and attempted to roll out of turn and recover the aircraft while passing the ship at low altitude. The aircraft impacted the water wings level. The AMB would conclude that power available was exceeded by power required. The aircraft commander was thrown from the aircraft, still strapped in his seat. The aircrewman, sitting in the door strapped in a gunner's belt, does not know how deep the aircraft sank before he could get out. The force of the crash was such that the aircrewman bit off part of his tongue, cracked some teeth, and was cut in several places while egressing the aircraft. When he broke the surface, he saw the aircraft commander sinking, and despite his injuries, he swam over, released the pilot from his seat, inflated his survival vest, and rendered immediate first aid until the ship's motor whaleboat arrived. He never saw the copilot. The AMB did not find any evidence of mechanical or structural failure. The crash was attributed to pilot error.

We had a memorial service at the Base Chapel and a flyover by one of our aircraft as the congregation was leaving the service. The Barile family was looking for answers as to what had happened, and we could not talk about the AMB or the Judge Advocate

General's investigation. I told Mr. Barile that the aircrew was not doing any maneuver that I had not done in my naval career. I told him that you expanded your parameters as you became more comfortable and confident in your ability to fly the aircraft. I compared it to learning to drive a car. When you first started out, you were very conservative about how you drove the car. You obeyed all the speed limits and observed the proper distance between vehicles. As you gained experience, you drove differently as you learned what the vehicle could do and how much you could handle. I told him his son Ralph was at the beginning stage of being a naval aviator and that, over time, he would become more comfortable in flying the aircraft. I said that part of being a naval aviator involved taking the aircraft to its limits, or even a little beyond so that you could survive in combat or in an emergency where you are forced to execute maneuvers that you would not normally do.

Ralph's OINC was a Weapons and Tactics Instructor who had been trained at Marine Weapons and Tactics Squadron One. He had been trained to do the maneuvers he was executing the day of the mishap. It was only after the accident that we learned from some of the pilots in the squadron that they had been uncomfortable flying with this aircraft commander. They felt that he had exceeded the parameters for the maneuvers and would do the maneuvers without letting the copilot or aircrewman know in advance. These pilots had not gone to the Safety Officer, the Skipper, or me to tell us they were scared to fly with this pilot. I stood up at the next All Officers Meeting and told the Wardroom I would never forgive them for not telling me about this unsafe pilot. I said I would have confronted him about it and told him to change his ways. Whether he would have and whether Ralph Barile would still be alive was unknown, but at least I would have known I tried my best. I hated them for not giving me that opportunity. We finished that AOM by discussing the notion of who the next pilot in the squadron to have an accident was and how we would be able to keep that from happening.

The family wanted to know if the Navy would attempt to retrieve the wreckage. Based on the depth of water in which the accident occurred, and the fact that we did not believe there was anything wrong with the aircraft, the Navy decided not to retrieve the wreckage.

Following the accident, we had several briefings where we brought the members of the squadron and their families up to speed on what happens and what they should expect from the squadron when there is an accident. Over my career, I adopted an attitude that you should face the monster in the closet rather than trying to ignore it and hoping it would go away. If you are prepared to accept the worst thing that can happen, it makes dealing with anything less than that more manageable.

A couple of months after the fatal accident, we got a report that one of our aircraft might have gone down near Kauai. The aircraft had been on a scheduled mission at the Pacific Missile Range Facility at Barking Sands to do some ASW work and was on its way home when a fisherman reported seeing the aircraft go into the water along the southern coast of Kauai. Very soon after that, we received a message that the aircrew had been recovered and were all in good condition. While the Skipper was dealing with the pilot's families, I called the AW's wife. She was upset with the news and wanted to know her husband's condition. I said we were only told they were not injured and that we expected them to return to the squadron that evening. I referred to one of our briefings and said, "remember when we told you that if the aircrew are killed or seriously injured, we will come to your home?" She calmed down a bit, and I promised to get back to her as soon as we got more information or had a definite time on when they would be back in the squadron. Her husband was back at home that evening with no injuries.

The subsequent mishap board determined that the cause of the accident was the departure of one of the flaps on the aircraft's main rotor blade. This caused severe vibrations in the aircraft, and the aircraft commander felt that the aircraft was becoming uncontrollable. He decided to descend and put the aircraft into the water about a mile offshore while he still had some control. Everyone got out of the aircraft without injury.

While I had been away at the War College, the squadron had an accident for which I was at least partly responsible. As an OINC and Maintenance Officer, I used to say that you should never pass up an opportunity to get gas or go to the bathroom when on a flight.

One day a deployed detachment was making a logistic run from their destroyer to an oiler to deliver some cargo. When they landed, the OINC, an East Coast guy, said he remembered my admonition and asked to have the aircraft refueled. They did not need the gas to make the return flight to their ship. They topped off on JP-5 fuel and were about halfway back to their ship when they began losing power in both their engines. They went to emergency throttle to manually control the fuel, i.e., to program maximum fuel to the engines, but they continued to lose power. They notified their ship and requested the ship close its position at flank speed while they jettisoned everything they could from the aircraft. The engines continued to lose power, and the rotor rpm began to decrease. On the H-2, when you reached a certain rotor rpm, the #2 generator would drop off the line. It was nighttime, and the aircraft commander wanted to ditch with power and lights on in the aircraft. He notified the ship, and the aircrew completed their ditching procedures. The aircraft commander lowered the aircraft into the water and rolled it to the right, the side on which he was sitting.

After all motion stopped, he exited the aircraft, and when he broke the surface, he could see the copilot and the aircrewman. They linked their survival vests together and waited for their ship. At one point, they were kicking one another, and no one wanted to say anything because they feared it was a shark. Finally, one of the crew asked whether the others had felt anything. They all said yes, stopped kicking, and the "phantom shark" never came back.

The Accident Mishap Board determined that the aircraft had received contaminated fuel from the oiler, and that was why the engines lost power and would have eventually flamed out. I modified my guidance to never pass up the opportunity to get "good" gas or go to the bathroom.

As time went by, I became more frustrated with Skipper's style of leadership. The squadron was slipping, and he did not feel the same sense of urgency that I did to stop the decline. Our discipline and professionalism were not where they should have been. I was conflicted about how hard to push things or whether to just wait it out until I became Skipper. I was becoming a disloyal XO. At department head meetings, I let it be known that certain policies and procedures would be changing when I assumed command.

Time seemed to fly, and it was not long before I received a call from the Detailer on who my prospective XO would be. A couple of names were mentioned, and I said I did not believe that one of them would be a good fit based on family concerns and how difficult it was to adjust to life in Hawaii. If you had kids or dogs, the transition to life on Oahu could be difficult. The Hawaiian school system was designed to produce graduates who could perform well in that environment. Unfortunately, most military kids would not remain in the "hang loose" environment of the islands. They would be returning to places in the states like Fairfax, VA,

Newport, RI, Virginia Beach, VA, or San Diego, CA, where the academics were much more difficult, and a lot more was expected from the students. If you had dogs, they had to be put in quarantine on the island. This added additional expense and emotional trauma for families coming to the islands for the first time. Although I did not know or have ever met him, Commander Dave Taylor seemed like a good fit at XO. He had been stationed in HSL-37 before and had a solid reputation in the SH-2F community. I looked forward to welcoming him and his family back to the Easyriders.

CHAPTER XIII

SKIPPER AT LAST

(NOBODY SAW THIS COMING)

HSL-37 EASYRIDERS

JUNE 1990 – JUNE 1991

The day of the change of command was just another day in paradise. There is no place in the world to live like Hawaii, and the family and I enjoyed it tremendously. Mom and Dad were there along with my best and closest lifelong friend Pat Foy, former squadron mates, and Bill and Lois Harkins from our college days. The trade winds were blowing, and the sun was out as I relieved Skipper Steele and stepped to the podium. The speech I gave outlined what I expected of the Squadron and follows.

"Admiral Schriefer (Commander Anti-Submarine Warfare Wings Pacific), Captain Milioti (Commanding Officer NAS Barbers Point), Patrol Wing TWO COs and XO's, our Hangar Mates VC-1, Ladies and Gentlemen, and the superb professionals of HSL-37, the Easyriders.

Today is truly a great day. It is the realization of a personal goal of some seventeen years, aviation command.

I understand the nature of our respective duties here today. Mine is to make my speech succinct. Yours is to listen and for some of you to time the length of my succinct remarks. I promise to be brief if you promise to listen. First, I would like to thank Admiral

Schriefer for taking time out of his schedule to participate in our change of command. I know that with the changing fiscal guidelines coming out of AIRPAC these days, he does not want to spend too much time out of the office, or he will return to find out they have cut all his money.

Captain Milioti, I thank you for your support in enabling us to complete our mission and the beautiful setting we enjoy this morning.

Skipper Steele, you have provided the solid base upon which to build. I have benefitted greatly from your tolerance and patience. I have admired your performance under pressure during the Squadron's darkest hour last November. I wish you, Chris, and the kids the best as you depart for the Naval War College in Newport. I would like to pass on two important things I learned during my time there. One, it is only a lot of reading if you do it. Two, showing up is 90% of the grade.

Dave and Caroline Taylor, Mathew, Katie, and Zach, welcome back to paradise. I know the wealth of experience you bring will benefit all of us in the year ahead.

As I mentioned a few moments ago, this day represents the culmination of my personal and professional goals over the last seventeen years. But, in a very real sense, I share this day with a former Navy Chief Aviation Ordnanceman, my Dad. As everyone knows, the center of the universe in an Irish house is the kitchen, where strong conversation and liquid spirits are freely exchanged. It was there I first heard of the accounts of the fast carrier task forces of World War Two in the Pacific. The exploits of Task Forces 38 and 58 under Admirals Halsey and Spruance were indelibly burned into my brain as I listened to my father and my uncles recount their experiences. There was also a realization that the finest thing in life was to be a part of Naval Aviation. While most of you think of the USS San Jacinto as the ship on which President George Bush served, to me, it will always be my Dad's ship. And while a shrinking post-World War Two Navy and a growing family forced my Dad to opt for better economic conditions on the outside, I somehow felt he always wanted to have completed a naval career. Dad, this day is both of ours.

Mom, from you, I learned the meaning of grace under pressure and having a sense of humor. Raising four sons and a daughter on a policeman's salary was not easy. Some of the Navy wives in the audience today could learn much from you about how to cope with being married to a husband who faces dangerous situations every day. But no matter what happened, your ability to look life in the eye and laugh at adversity is something I have always admired and tried to emulate.

AUTHOR WITH MOTHER AND FATHER AT HSL-37 CHANGE OF COMMAND

There are some other people who I must thank. Dave Youngblood was my first OINC, and he tactfully explained amidst his laughter that you do not start every sentence in a message to the Squadron with ADTAKE, addressee take action. He also wisely restrained a young lieutenant from doing touch and goes on a stretch of desert road running between Damietta and Port Said in Egypt.

Former Easyrider CO's Captain Mike Putnam and Commander Bill Fetzer. Captain Putnam for giving me the best detachment a man could ever want, the Bounty Hunters of Detachment Four. And the latitude to run my own show, even if I was from the East Coast.

Commander Fetzer, for having the confidence in me to give me the best job I ever had, maintenance officer of this Squadron. His support and guidance in the modification of the first four "Attack Wolves" in the summer of 1987 was outstanding. Admiral, no LAMPS outfit could have done it better or more professionally.

To some friends who have come from far away to share this special day. Ed Driscoll, who today masquerades as a Vice President for Kaman Aerospace, but who used to earn an honest dollar as my OINC. Ed invited me to consider many things when I worked for him. He taught me two important lessons, take care of your people, and they will take care of you, and do not accept the status quo. Always try to change things for the better.

Art and Terri Gallo, from brand "X," better known as HSL-33, and their handsome and already balding son Art. As air station pilots in Bermuda on our initial tour, neither of us could have imagined that the Navy would ever be crazy enough to select us for command. As with most Ensigns, we always had the answers for everything. Art, it is time for us to put up or shut up.

Julie Barile, who came all the way from New Jersey to share this day with us, and her husband Ralph, who will always remain on deployment.

To my wife Linda, the love of my life. For the understanding and support you have always provided. Through back-to-back sea duty tours and deployments, through the moves from DC to Hawaii, to Newport, and back to Hawaii over the last four years. For the tolerance of a husband who seems to love flying and the sea more than he does you. Without you, none of this would have happened. Your support has allowed me to pursue the goal we both realize today.

To the girls, you brighten every day of your Dad's life.

And to the Easyriders, I stand here today because of you. There is no doubt in my mind that the efforts of your predecessors on Det 4 and in the Maintenance Department made it possible for me to stand up here today. Folks like Chief Cornell, Petty Officers Neuwirth and Sequin, Gunner Rosario, Lieutenant Jim Gilchrist, Master Chief Herring, and so many others who I cannot possibly name. They helped to put me on this podium. I want to outline for you our goals for the next year.

They are only two, combat readiness and a clean sweep.

Combat Readiness to me, is each member of this Squadron realizing their fullest potential. To be the most highly trained and proficient aircrews. The most professional and dedicated maintenance personnel, an administrative support team that is second to none.

To the Khaki and Leading Petty Officers, take care of your people, and they will take care of you. Ensure that we, you and I, do everything possible to educate, train, promote, and recognize the truly outstanding individuals among us. Be professionals, and know all facets of your job. Take and accept responsibility for your actions, both when right and wrong. Take the time to do the job right the first time.

A clean sweep. Admiral, you can tell our sister squadrons in San Diego, if that spy Commander Gallo doesn't, that the Easyriders are on the move. Our goal is to win the Battle "E," CNO

Safety Award, ISBELL Trophy for ASW excellence, the Top Torpedo award, and to be the Golden Anchor nominee. To do what our predecessors in HSL-37 did in 1987. We seek these awards not for the awards themselves but for what they represent. That each member of this Squadron has realized their fullest potential by ensuring combat readiness. We know that we probably won't achieve this clean sweep, but we can do no less than give it our all.

Why do we need to realize our fullest combat readiness?

Because of the unique operations required on a daily basis from LAMPS detachments, I know my surface brethren understand this. The USS Samuel B. Roberts and USS Stark serve as reminders that we go in harm's way daily without the luxury of sophisticated support facilities or the protection of an Airwing. LAMPS operations off Lebanon, Libya, Nicaragua, and in the Persian Gulf have demonstrated the versatility of the LAMPS/Ship team. They also remind us that in these days of the so-called collapse of communism, our duties bring us into conflict more often in the third world, and are more likely to do so in the future.

We cannot allow those who say that peace is breaking out all over to lull us into a false sense of security. The philosopher Santana said those who fail to learn from the mistakes of history are doomed to repeat them. We have seen the ebb and flow of the budget battles before. We have seen our predecessors called upon to act as instruments of American foreign policy all over the world. We must remind ourselves that the call will come again, although we do not know where or when. What we do know is that we, the Easyriders of HSL-37, must be ready, and we will."

The Change of Command was in June of 1990. In August, Saddam Hussein would invade Kuwait, and the Easyriders would go to war.

Just prior to the change of command, the Squadron was notified that we would be using a new system for supply and maintenance, Aviation Depot Level Repairable or (AVDLR). Previously if you needed a part, you requisitioned it from supply and sent the old one to be repaired. What it cost or how long it took to fix it was not the Squadron's concern; it was up to the Naval Air Station Supply and the Aircraft Intermediate Maintenance Department (AIMD). Under this new system, you were given a maintenance budget and were required to manage the removal and replacement of major aircraft components by staying within that budget. This forced the Squadron to prioritize maintenance and management among the deployed detachments and the home guard. You now had to check to see if you had money to pay for the parts that you wanted to remove and replace. You could change a main gearbox or four main rotor blades, but you did not have the money to do both. When we were briefed on the system, Skipper Steele looked and me and said, "good luck with that!"

So now I became best friends with the NAS Barber's Point Comptroller, the Supply Officer, and the head of AIMD. I felt a lot like the cartoon character Wimpy from the Popeye cartoons. "I'd gladly pay you next Tuesday for a hamburger today," but in my case, it was "can you give me that part on credit, and I'll pay you back next month?" Our maintenance officer LCDR Bob Asmus did a fantastic job balancing all the requirements to keep us up and operating. We were now forced to prioritize maintenance. The other thing we learned was that my maintenance budget would be controlled by the P-3 Wing located on Barbers Point and not ASWWINGSPAC in San Diego. I think ASWWINGSPAC liked telling me there was nothing they could do for me and that we just needed to work out our issues with the P-3 folks. The thing I did not tell the Wing in San Diego was that the P-3 Wing in Hawaii was great to work with and that the dollar amounts for our budget were much smaller than theirs and so they tended to give us the benefit of the doubt when we would ask for more money. Parts for the H-2 also cost a lot less than for a P-3, so it did not seem like we were asking for that much money.

Our flight hour money still came from San Diego, and we had no problems with that. A squadron's reputation was based on how well you executed the flight hour program. That meant exceeding your flight hour allocation. It seemed like the Skippers who tried to manage the system honestly were the ones who suffered the most. It was a lot like the story about the grasshopper and the ant.

The honest Skipper Ant would sit down with his folks and try to forecast how many hours the Squadron would fly over the course of a fiscal year. That would be based on pilot training requirements, proficiency, and operational requirements. You knew how many pilots you had and their annual hour requirements and then projected what the deployed detachments would fly. A good Skipper would consider all that and submit the squadron requirement to the wing. He would then manage the figure to ensure that the Squadron burned the flight hours as projected and flew the last flight hour on the last day of the fiscal year.

The average Skipper Grasshopper would produce an approximation and try to fly those flight hours as soon as possible. He would get to the fourth quarter and then go to the Wing and say, we will have to park the birds for the next three months because I have run out of flight hours. So, the Wing would look at the good Skipper Ant and say, you have not flown all your hours, so we are going to take some of yours and give them to your fellow Skipper Grasshopper. So, the man who was the better manager, Skipper Ant, had to rescue the inefficient Skipper Grasshopper by giving him flight hours that he would burn through and then ask the Wing for even more flight hours. The only reason the system worked is because in the last part of the fourth quarter, there were inevitably squadrons that could not execute their projected program and therefore had hours to return. Also, the cost per flight hour meant that when a fixed-wing squadron, say one flying F-14s, turned in a flight hour, you could fly a couple of H-2s all day long on the same amount of gas.

We began to experience trouble with our trim strut actuators. These actuators would electro/mechanically position the struts controlling the cyclic or rudder pedals. In some cases, they did this automatically in response to the pilot's control inputs, or they could be trimmed manually. Some were beginning to fail, and when they did, they automatically went to the extreme control position without any pilot input. This failure could be quite dramatic, and depending upon when it occurred, it was unsafe and dangerous.

The situation came to a head when the Operations Officer was out flying a hop and had a trim strut failure in the cyclic that went to the extreme position. He said that he could barely get the aircraft on the ground. Operations and the Safety Officer came into the office and said we had to do something about the situation before someone was injured. Two different companies manufactured the Trim Struts. One of the companies had not had any trim strut failures, while the other had quite a few. We did an inventory on our home guard aircraft and requested the deployed detachments do the same. About half of the trim struts were the so-called "bad ones." We stripped the home guard to send the "good trim struts" out to the deployed detachments and cautioned them to brief this situation and their response before every flight. That meant that we only had a couple of aircraft at home that we could use for training and proficiency.

We notified the Wing and asked for guidance on ordering the replacement trim struts. We could say all our aircraft were down for unscheduled maintenance due to supply, and then we would get priority on the parts but would not be able to fly. The Wing directed us to order the trim struts as scheduled maintenance, which would mean that it would take a longer time to get them. The East Coast Squadrons had all ordered their trim struts while we were waiting on the West Coast wing to tell us what to do, so we went to the bottom of the list for delivery. At the time, I considered sending an officer TAD with the "good" trim struts out to the deployed detachments so that they would be hand-delivered rather than sending them through the mail. As it turned out, the trim struts the detachment ordered arrived before the ones through the mail, even though it took a couple of months.

One of the other things the Wing said was, "don't send any messages into the reporting system that changes the status of your aircraft based on the lack of available trim struts." I did not concur with this as I had been raised, by Ed Driscoll, that you were supposed to accurately reflect the status of your aircraft and that the only way to document the need for repair parts was through accurate messages to the supply chain. Inevitably when the maintainers said that a particular part was malfunctioning, the supply folks would ask for the documentation where you had shown that a particular part was failing at a greater-than-expected rate. Unless you documented the equipment failures, the supply system would not stock the system with repair parts to account for the failure rate. The result was you were out of the part, and supply said they did not show any demand. But after I had expressed my misgivings to the Wing, I instructed the Squadron that we would not be sending out any messages about trim struts. There is always someone who does not get the word. In this case, one of the detachments, who was home in the Squadron at Barbers Point, sent out a message saying that they would be going down in a week for a trim strut.

My first indication of things not going well was when I received a call on Saturday from the Admiral while I was out on the golf course. The Admiral wanted to know why we had sent a message on trim struts when we had been instructed not to do so. I was adamant that there must be some mistake on the Wing's part because I had made it clear to the Squadron that we would not be sending any messages out on trim struts. The Admiral indicated in extraordinarily strong language that I did not know what was going on in my Squadron and I better straighten this mess out quickly. I called the Squadron and found out about the detachment's trim strut message. Then I called the Admiral back and profusely apologized for causing the Wing any trouble. He was less than satisfied.

That Monday, I called the offending OINC and his detachment maintenance officer into my office and read them the riot act. I revoked their authority to release messages, questioned their competence, and said something about wishing I had a yardarm to hang them from. I threw them out of the office and instructed the XO to revoke all the detachments that were in the Squadron's message releasing authority. He nodded. About a week later, I thought better of what I had ordered. The dets needed to learn how to do their messages properly before they left on deployment. I told Dave Taylor to rescind the restriction on detachment message release authority. He smiled, and I asked him why. He said, "Skipper, I knew that you really didn't mean what you said about the message release authority and that you would come around once you cooled off. So, I never told the detachments to stop sending messages. I just had them bring them to me, and I released them after they were reviewed." It was not the last time that the XO would save me from myself.

Some years later, that detachment maintenance officer became the Skipper of the Easyriders. Linda and I were visiting Oahu, and he invited us to the Squadron and then later to his house for dinner. We were talking about what it is like to be in command and how you cannot really judge the impact you are having on your people at the time when you are in command. He asked me if I remembered the trim strut message incident. I said not really. He had been the maintenance officer on the detachment that had sent out the message. He told me that when he left my office that day, he felt his naval career was over, and he would be relieved. He carried that incident around with him for all those years, and I had forgotten about it completely.

The other incident that got me seriously crossways with ASWWINGSPAC involved a visit by Commander Naval Air Force U.S. Pacific Fleet (COMNAVAIRPAC) to the Squadron. COMNAVAIRPAC was scheduled for a visit to Barbers Point but was not supposed to visit the Easyriders. He was scheduled to visit VC-1, a composite squadron that flew A-4s and H-53s that shared the hangar with us. We spruced up our spaces just in case the Admiral decided to come over, but we were doing business as usual, which meant maintaining aircraft and flying a full flight schedule. We watched the Admiral arrive and go into VC-1. Sometime later, we received word that the Admiral was coming over to our spaces and wanted to meet with the XO and me.

COMNAVAIRPAC wanted to talk with us about the fatal accident that occurred when I was XO. The Admiral wanted to know what our screening process for Officers in Charge was, their training, and how we monitored them. He was especially interested in the background of the OINC, who had been involved in the fatal accident. We told the Admiral that the OINC had come to us from a tour in the FRS, where he had been the Quality Assurance Officer and one of their top instructors. He had also been through the MAWTS evasive maneuvering course. The Admiral felt the system had failed to weed out a bad pilot. We pushed back that there was no way to know for sure how an OINC would perform until they had been to sea. We monitored them closely during workups, and ASWINGPAC would assign a safety observer to make sure that the detachment was operating properly. There was not much else we could do.

The Admiral also wanted to know how we felt about teaching Evasive Maneuvering to break the lock on a fire control radar. This involved using extreme angle of bank and simultaneously reducing your altitude. It was like a 180-degree autorotation on steroids. AIRPAC was weighing the risks in teaching and doing the maneuver vice the likelihood that you would have to do it operationally. The concern was that you might lose an aircraft while training on the maneuver. We said it was critical that we do the training and that we had set up a syllabus with PMRF Barking Sands using their fire control radars that allowed us to perform the training locally rather than having to send aircrews back to San Diego. He was still not convinced that the LAMPS community had the right answers to qualify our OINCs or break lock.

Then the Admiral asked if he could speak to our pilots. I said that it would be difficult to round them up as we were running a full flight schedule but that we would assemble those that were available in the wardroom. He opened with some general comments about naval aviation and said he had heard good things about the Squadron. He then asked if there were any questions. No one raised their hand. Finally, he turned to one of our Lieutenant Commanders and said, "Commander, I'm sure you have a question for me." Lieutenant Commander Deke Kennison hesitated, so the Admiral asked him again, "What is on your mind?" This time Deke responded, "Admiral, when are we going to properly man, train, and equip for the Combat Search and Rescue Mission (CSAR)? Every LAMPS detachment is asked about their capability to do CSAR at some point during a cruise. We waste a lot of time and energy in producing temporary Ad Hoc solutions. What we need to do is arm our helicopters and train on the proper tactics, techniques, and procedures." You could have heard a pin drop in the room. The Admiral looked around and fumbled to produce an answer. He started talking about the JVX, which would become the V-22 Osprey, as the answer to the CSAR technical requirements. He did not mention that aircraft was years away from being developed and that, in the meantime, we would continue to have to come up with Ad Hoc solutions. But the Junior Officers in the room knew that the Senior Aviation Admiral in the Pacific did not have a clue as to what the LAMPS community did for a living on deployment and was not interested unless it involved the Fighter/Attack communities. The Admiral quickly left, and I knew that the visit had been a disaster.

I had to give my boss COMASWWINGSPAC a heads up on the COMNAVAIRPAC visit, and I went back to my office and picked up the phone. COMASWWINGSPAC was not available, so I asked to talk to the Chief of Staff. I outlined the visit and said that it was clear that AIRPAC had not been properly briefed by his staff on our Squadron and missions. I asked the Chief of Staff to brief the Admiral and said I would be happy to provide any additional details if necessary. It was a big mistake on my part. I should have asked when the Admiral would be available and made sure to personally give him a detailed debrief.

I learned later that AIRPAC ambushed ASWWINGSPAC about the "whiners" in HSL-37 and what a terrible squadron it was. The two Admirals did not get along especially well to begin with, and I had provided AIRPAC with ammunition to "wire brush" my boss. The next time I was in San Diego for a wing meeting, I had a one-on-one with ASWWINGSPAC where he told me how disappointed he was in me and the way I was running the Squadron. For a few minutes, I really felt that he was going to relieve me because of the AIRPAC visit and the trim strut message. I found out later that when the Admiral and I left the meeting room, one of my fellow Skippers remarked that that was the last time we would see Jack as Skipper of HSL-37.

When Saddam Hussein invaded Kuwait, the U.S. Navy immediately dispatched the USS INDEPENDENCE Battle Group to the area. We had a detachment embarked on one of the frigates in the Battle Group. So, we were naturally concerned about them and how they would be operating. At the same time, we received a message from the Wing asking how many detachments you can surge, if necessary, to the Persian Gulf. The XO and I met with the Department Heads and started to do some calculations on forming additional detachments. We took it all the way to both the XO and I deploying. Our respective detachments would have the most junior pilots and maintenance folks left in the Squadron after the other detachments had been formed, as well as the least desirable aircraft. As always, the limiting factors were support equipment (yellow gear) and logistic pack-up kits. We did not have enough tools or spare parts to go around. It does not do a lot of good to deploy, and when something breaks, you do not have the required replacement part or the tools or calibration gear necessary to do the repair. Thankfully, we never got that far and were able to deploy fully trained and equipped detachments.

During this planning, we had an all-officers meeting in the wardroom. I reminded the pilots that being a naval aviator was not just about liberty ports and having a lot of fun. Occasionally, like now, the U.S. Navy is going to call in its marker and ask you to go in harm's way. I used a phrase I had heard years before about Coast Guard helo pilots (our fellow naval aviators), who traditionally were named the rotary Wing aircrews of the year for their daring rescues under the most dangerous conditions. I said, "You have to go out. You don't have to come back." I told them they needed to think seriously about the prospect of going into combat and prepare themselves accordingly. If they did not think they could manage it, now was the time to come forward.

There was a minimum of information coming back from the detachments as they were deployed, but several of them cross-decked to other ships during the war. At one point, we had two detachments onboard the same ship. The junior OINC had deployed with the Spruance class destroyer, while the senior OINC was coming from a Knox class frigate. Normally the senior man would oversee both detachments, but I felt that the OINC that had established a relationship with the ship should remain in charge. It was tough to send that message, but the two were professionals and worked it out.

The LAMPS community was unique in that the CO and XO did not deploy but stayed at home and managed the detachment's formation, training, qualification, certification, and deployment. During the Gulf War, this was difficult as the XO and I were the most experienced aviators in the Squadron. Waving goodbye was not in our DNA. In Vietnam, the COs and XOs were normally in the first aircraft launched on a dangerous mission. I had read stories about junior officers being physically yanked out of their aircraft while sitting on the catapult and replaced by a more senior pilot. The thing that had drawn me to the LAMPS community in the first place was that I could be that guy as a Lieutenant, making the call about who and how we would conduct combat operations, rather than having to wait until I was the Skipper of a squadron. Ironically, I had been an OINC too early to be able to take part in the only war that would occur while I was operationally flying. I knew that our OINCs were professionals, and they could manage the challenges of potential combat. I also knew that the kids would grow up fast in that environment.

Between the challenges of AVDLR and the Gulf War, there was not a lot of flight time left for the home guard, especially the CO, who is at the bottom as far as required flight time. I enjoyed flying with the JOs and tried to do so as often as possible. I especially wanted to go on the boat with them for their day and night deck landing qualifications as often as possible. Like flying under actual instrument conditions, you can learn a lot about an aviator when he is trying to get aboard the ship at night. I also liked doing their NATOP and Instrument checks whenever possible. When I was doing their instrument check, I would give them three options: (1) you can file a flight plan to one of the other islands; (2) you can do two hours of basic instrument work; or (3) we can go tower to tower over to Marine Corps Air Station Kaneohe, shoot a couple of approaches, and call it a day. The whole thing would take about an hour at most. Most folks chose option (3).

What I knew was that going to Kaneohe guaranteed that you would have actual instrument time because of the clouds that formed over the mountain range that splits Oahu, as well as the likelihood that you would encounter actual instrument conditions while shooting the approach into Kaneohe. You can fake a lot when you have a hood on to simulate instrument conditions. It is apparent very quickly when you are in "the goo" (remember that highly technical term denoting no visible horizon), and you cannot cheat by looking through the cockpit floor windows whether you have been practicing your instrument flying regularly. I had previously done these instrument checks with the JOs when I was the maintenance officer in the Squadron. I would not make them wear the hood because I knew we would go into instrument conditions, and I would be able to tell how good they were at instrument flying.

Back when I was the Maintenance Officer in the Squadron, it was time for my instrument check. I would be flying with Mike Carnes, the Operations Officer. He asked me what I wanted to do, and I said tower to tower to Kaneohe. He grinned and said let's go for it. He asked me to put a hood on right after we took off. He gave me an emergency to see if I would continue the flight or return to Barbers Point. After I said I would return to Barbers Point, he waited a few seconds and then kicked off the Automatic Stabilization Equipment (ASE). This meant it was more difficult to fly the aircraft while trying to concentrate on my instrument procedures. He doubled my workload. After a few minutes, he came on the ICS, said he had vertigo, and pulled the hood off my helmet. I looked up from the instrument panel, and we were solidly in the goo. His action in pulling off my hood had also induced a slight vertigo in me. My reaction over the ICS was something to the effect of what the f--- was that for. I quickly got it together, and the rest of the flight was uneventful. I am sure the JO's asked him to give me a taste of my own medicine.

One of the disappointing things about being the Skipper was that I did not get to fly as much as I would have liked. The priority for flight hours was the detachments and building the flight experience of the junior pilots at home. The old farts, like me, got the minimum necessary to maintain our qualifications. I had hoped to get a Kaman plaque for 3,000 hours in the aircraft but wound up with about 2,700. Besides, I realized that I was getting paid to run the Squadron effectively, not lead it in flight time.

Ironically, one of the most challenging aspects of being in command was managing the information and expectations of the spouses. We held events in the Squadron, pre-deployment briefings, and even had the detachments over to our house on the beach to provide them with as much information as possible about their deployment and the challenges they would be facing. It becomes obvious very quickly who will be able to meet the tremendous challenge of supporting a deployed spouse and those that will need additional help. We tried to be honest and upfront about how hard it would be, but some folks just did not want to believe it. At the conclusion of the Gulf War, we were trying to determine how to get one of the forward-deployed detachments home. There were at least three possibilities, and I had a meeting with the spouses in the wardroom to go over what they were. At the end of the briefing, I was asked which one I thought was the most likely. I gave my opinion, and of course, I was wrong. I heard later that one of the spouses was saying that I had lied to them about how the detachment would get home. Linda had been at the briefing and was disappointed that the spouse could say such a thing when I had gone out of my way to caveat my response.

About halfway through my tour as Skipper, the Wing wanted us to take over the Attack Wolves again. Those were the unique aircraft that had been modified while I was Maintenance Officer. I did not want to take them back. We would have to do an at-sea transfer, and our crews would have minimum training prior to deployment on the revised tactics, techniques, and procedures that were being used with the aircraft. The San Diego squadrons had all the recent experience, and I felt they should continue to operate the aircraft. But we would be deploying on the ships that would be replacing those on the station, so we accepted the Attack Wolves. It is a testament to the professionalism of our detachments that they were able to do the aircraft transfers, unique specific on-the-job training, and complete the deployment without a hitch.

Also, about this time, there were discussions about introducing the Mark III SH-60B Seahawks on Oahu to support the ships deploying out of Pearl Harbor. Rather than stand up a new LAMPS Mark III squadron, the Easyriders would become a composite squadron, made up of the Mark I SH-2F and the Mark III SH-60B. My initial efforts were to fight for the spaces in our hangar to house the new aircraft and personnel while starting the process of ensuring that the Barbers Point Commanding Officer, as well as his AIMD and Supply Officers, were aware of what was coming. We also talked with the CO of the SH-60 FRS in San Diego about getting our current and future XO to go through the SH-60 CO/XO training. CDR Dave Taylor and I had talked about how he would get to enjoy the benefits of the challenging work we had done to prepare the Squadron for the SH-60 transition. HSL-37 was the first Squadron to transition from the SH-2F to the SH-60B beginning in 1992.

It seemed like I had only been the Skipper for a few months, and the prospect of my next tour reared its ugly head. The normal career path for a LAMPS skipper following command would be to go to sea as either the Operations Officer or Air Operations Officer on an amphibious ship. These were the billets that made you a front-runner for selection to Captain, Major Command, and flag rank. I did not want to do any of those jobs. The idea of being chewed out by the Captain of the ship because some 18-year-old's haircut was out of regulation just did not appeal to me. My DC tour had broadened my horizon. I wanted to work where the big picture strategic policy decisions were being made and where I could have input into them, not at the tactical level on an amphib. So, I threw my hat into the ring for a fellowship at one of the prestigious think tanks. COMASWWINGSPAC was not thrilled with my decision but said he would support it. To me, any thought of getting a good FITREP was out the window, and doing a fellowship would help smooth the transition to life after the Navy, which was becoming increasingly likely for me.

I did not make the first cut for a fellowship. My detailer said I was on the alternate list, and it was hard to tell if I would get a spot. So, we talked about me going to one of the amphibs for my follow-on tour. We did not specify which one, just that we would talk about it later. I called the detailer who oversaw the fellowships, and he assured me that I was the first alternate, that at least one spot would be opening, and that I would get a fellowship.

Imagine my surprise when my detailer called and said, "you are going to Supreme Allied Command Atlantic (SACLANT), a North Atlantic Treaty Organization (NATO) Command in Norfolk." When I asked why he said, "it is a Joint Staff Officer (JSO) billet, and you are one of the few qualified JSOs in the Navy that are rolling in that time frame, so you will be going there."

Suddenly, going to the amphib never looked so good. As far as being competitive for Captain or Major Command, I had once heard another officer say that going on a NATO tour was a little bit better than being on leave, but not by much. The detailer then asked me how I got to be a JSO when I had never served in a joint billet.

Joking, I said that I had once qualified Army aviators to land onboard a Coast Guard ship, and maybe that counted. That was one of the highlights of my time as CO. I got to fly in an OH-58 Kiowa with a senior Army Warrant Officer who had been in the 160th Special Operations Aviation Regiment (The Nightstalkers). I talked him through the approach to the ship, and as I expected, he was on the money the whole way down. I was not supposed to fly the aircraft, but after he had completed the required number of approaches and landings to qualify, he asked me if I wanted to try a couple. I jumped on the controls and made a couple of passes. The Navy's primary trainer, the TH-57 Jet Ranger, was like the Kiowa. The Kiowa was a lot more sensitive than the H-2, and I had to remind myself not to "plant" it on the deck. I also had the opportunity to do shore-based training for an AH-1 Cobra pilot but did not get to take him to the boat.

Anyway, I told the detailer that the OPNAV tour in OP-906 working on issues affecting the Navy and OSD, plus the master's degrees from Georgetown and the Naval War College, along with the Joint Professional Military Education I took at the Naval War College, must have been enough. When I asked if someone else could go in my place, he laughed and said that I would be going to a "joint duty critical" billet, and those could only be filled by a JSO like me.

The day for the change of command finally arrived. I felt a tremendous sense of pride in what our folks had accomplished. I knew that the Squadron was in a better place than it had been when I had taken command, much more professional, seasoned, and positioned for success. The list of accomplishments of the Squadron over my time in command included: the introduction of low-level flight training on Oahu using Army ranges; an increase in training readiness from 87-91%; two detachments deployed in support of DESERT SHIELD/DESERT STORM that flew over 1,023 hours; an increase in overall flight time by 7% to 6,447 hours; winner of the Top Torpedo Award for LAMPS MK 1 Squadrons on the West Coast; three detachments deployed on short notice to support Counter Narcotics operations off Central America; institution of break lock training using the Electronic Warfare Range at the Pacific Missile Range Facility, Barking Sands; executed 99% of our flight hours despite a 75% cut in our aviation operational maintenance budget; and formulated plans for the introduction of the SH-60B into the Squadron including aircraft, facilities, material, and personnel support.

On the day of the change of command, we met with the Admiral and went over the scheduled ceremony. He had not briefed me on my FITREP, and I was wondering when he would do so. We left the office, went to the change of command, and adjourned to the Officer's Club for the reception. I still had not received my FITREP. As the Admiral was leaving, I walked him out to his vehicle, and he handed me an envelope with my FITREP inside. I figured that it must be good, or he would have discussed it with me. My previous FITREP from the Admiral had been for only two months. He could have given me a not observed FITREP because I had been in command for less than ninety days. Instead, I was ranked 3 of 3 with all the right words, but with what I would have termed a pack minus FITREP. As mentioned earlier, Pack Plus meant you were doing very well. Pack meant that you were a standard Mark I, Mod 0 Naval Officer, and Pack Minus meant that you should be considering another career outside of the Navy. My FITREP for ten months in command had all the right words, but there was no comparison to my fellow COs. That comparison is what selection boards would be looking for. The fact that there was not one indicated that the Admiral did not think I had done a particularly respectable job. But he also did not put a stake through my heart.

Based on the conversations I had with detailers and placement officers over the years, I developed a briefing for the junior officers in the Squadron. I described it as the track system. You were either fast track, on track, or off track. Your FITREP and the jobs you were offered indicated where you fit into the community and Navy hierarchy. If you were always number one or showed progression throughout a tour, then you were doing well, Fast Track. If you never made it out of the 5% category and were never recommended for accelerated promotion, then you were not doing well, Off Track. As a first tour aviator looking to roll to shore duty from Squadron, if you were fast track, you would go to the FRS as an instructor, Admiral's aide, or an intern position on the joint staff. If you were on track, then it could be as an instructor in the training command, post-graduate school, or as a recruiter. As off track or pack minus, you would be going to an Air Station or a non-flying job. I told them they could survive one pack minus tour, but two in succession meant that you would have a tough time making Lieutenant Commander. I tried to ensure they understood that performance was the key and that doing the best you could every day was the primary way to ensure success in their naval careers. I also told them that there was no guaranteed career path. They should not spend a lot of time trying to game the system on where your next set of orders should be. If all else failed, I said to look at my career path. Guys who start out at air stations, transition to a new helicopter community, volunteer for a DC tour, and switch coasts for their department head tours do not succeed professionally, but I had, and so could they.

CHAPTER XIV

DEEP SIXED TO NATO

SUPREME ALLIED COMMANDER ATLANTIC - SACLANT

JUNE 1991 – AUGUST 1994

I felt I had been discarded by the LAMPS community. My billet at SACLANT was the Crisis Management Staff Officer with responsibilities including Rules of Engagement, NATO Alert System, Potential Military Response Options in Crisis, and Exercise Planning.

I found out when I got to Norfolk that I would be replacing an officer who had been selected for early retirement. The fact that I was in a NATO billet and the officer I would be replacing was leaving the Navy made me wonder if I would make Captain. The Commander whose job I was taking was very well respected within NATO and had a great reputation. You knew you were respected by your peers in NATO when they referred to you as the expert. Commander Dave Nimmer was the expert when it came to the Rules of Engagement and the NATO Alert System. I would have my hands full trying to measure up to the standard he had established. During our turnover, he said that progress in NATO was slow and that getting any changes in Military Committee documents would be a significant achievement. What I could not have known was how much NATO would change over the next three years with the demise of the Soviet Union and the Alliance's involvement in the Balkans.

SACLANT was one of three Major NATO Commanders (MNCs), the other two being Supreme Allied Commander Europe (SACEUR) located at Mons, Belgium, and Allied Command Channel (ACCHAN) located at Northwood in the United Kingdom. Each of the MNCs had several Major Subordinate Commanders (MSCs) working for them. In the case of SACLANT, they were: Eastern Atlantic Area (EASTLANT, Northwood, UK), Western Atlantic Area (WESTLANT, Norfolk, VA), Iberian Atlantic Area (IBERLANT, Oeiras, Portugal), Striking Fleet Atlantic (STRIKFLTLANT, Norfolk, VA), and Submarines Allied Command Atlantic (SUBACTLANT, Norfolk, VA). SACLANT's job was to wrangle all these cats as well as communicate and coordinate with our fellow MNCs.

I would become what I referred to as the Maritime Rules Of Engagement (ROE) Meister or ROE Master. I was the Chairman of the MNC ROE Review Board. SACLANT was responsible for the Maritime Rules of Engagement. As such, if the ROE concerned maritime forces, I was the guy who was supposed to know their applicability. There were also Land Rules of Engagement and Air Rules of Engagement. The ROE experts at the MNCs at that time were all operators, not lawyers. The operators would produce the proposed changes to ROE, and then we would run them by the lawyers to see if they were legal. The Air and Land communities did not have very extensive ROE. Their feeling was that if you had to use force, we would be in World War III, and there would not be any need for the subtleties that existed in the maritime ROE. We were able to update the Military Committee (MC) - 192 Maritime ROE and began working on what we called Combined and Joint ROE. The idea was to have air, land, and maritime rules in one document. Our initial efforts eventually became MC 362/2 Joint and Combined Rules of Engagement.

ROE were always a sticking point during NATO exercises. There was always the question of whether the NATO ROE were consistent or contradicted the existing National ROE. During one exercise, I got SACLANT to release a message asking what national ROE were in effect during the exercise. We never got an answer. There were also things called "national caveats." A nation's forces operating under NATO could choose not to execute certain missions. You found this out when they refused to do something. One example was the use of tear gas. Many nations considered these to be chemical weapons and would not allow their troops to participate in operations where they might be used, such as in riot control. A nation might also restrict the operational area of its forces, limiting them to certain geographical or ethnic regions. These types of prohibitions always presented challenges for the NATO commander.

The other often debated topic was the inherent right of self-defense. Commanders would ask, how do I protect my unit if I do not have any authorized ROE? The answer was that any commander could protect his unit without having an authorized ROE if he were under attack (Hostile Act) or if they received indicators of an impending attack (Hostile Intent). In the maritime world, that consisted of indicators like being locked up by a fire control radar or if someone aimed a weapon at you in conjunction with the fire control radar. In short, if you observed the kinds of activities that indicated an impending attack, you did not have to wait for them to fire at you before you could respond. Nothing prevented a naval commander from exercising their inherent right of self-defense.

It was interesting to see the difference in how ROE was perceived by the different MNC representatives. My CINCHAN UK naval counterpart seemed to be heavily influenced by the British experiences in Northern Ireland. The UK thought of ROE from a law enforcement point of view, not a military one, and there were profound operational implications. The UK automatic assumption was that the results of any military operations would eventually wind up in a civilian legal court. Therefore the ROE had to satisfy "civilian law enforcement" kinds of considerations. The U.S. idea was that if you were complying with the "law of armed conflict," then you were not going to wind up in a civilian court. We were in the military, not law enforcement.

Simultaneously, we engaged in the development of the NATO Precautionary System to replace the NATO Alert System. Under these systems, there were measures agreed to in advance by the nations that they would automatically execute when requested to do so by the MNCs. They fell into several categories, and the nations tied their national plans to these measures. For example, if SACLANT declared measure XYZ, the U.S. had agreed to perform specific actions with specific forces. The idea was that in the case of Soviet aggression, NATO had to have procedures and actions agreed to in advance by the nations. There would be no negotiating whether a nation would perform them. So, any changes to the system required communication and coordination with all the NATO nations. These were time-consuming and sometimes frustrating, but we told ourselves that having all the nations agree to perform certain actions sent a powerful signal.

One of the things that would come out during the meeting to change the Alert System was how some nations suspected other nations of trying to effect significant changes under the guise of routine administrative action. I remember we were updating one measure, and the Greek representative proposed a change to the supporting reference document, which had been updated. All three MNC representatives stated that we supported the change and there was no operational significance, just ensuring that the reference document was current. We went around the table asking for national concurrence with the proposed change. When it got to the Turk, he said he could not approve the change.

The Chairman of the working group explained that all we were doing was updating the reference. Nothing changed in the measure. The Turk repeated that he could not agree to the change. The Chairman called for an early lunch, and after the other national representatives had left the room, the three MNC representatives went up to talk with the Turk. You had a U.S. Navy Commander (Moi), UK Royal Navy Commander, and a Canadian Air Force Lieutenant Colonel. We assured him there was no hidden agenda and that the Greeks were not trying to trick him. He said that he would contact the Turkish General Staff about accepting the change. This occurred on a Monday, and we were scheduled to adjourn on Friday. Every day the Chairman asked the Turk if he had received instructions from his government, and every day, he said he had not, and therefore Turkey could not agree to the change. We had given up hope of getting the revised measure approved. On Friday morning, the Chairman opened the session and asked if any of the representatives had any comments. The Turk called for the floor and said his nation could agree to the revised measure. I was amazed at the suspicion that some of the nations had for one another. I would see much more of it in my subsequent NATO tours.

My introduction to the world of exercise planning came about very quickly. NATO had a major Crisis Management Exercise, or CMX, every year. It involved the nations in NATO, the North Atlantic Council, NATO Military Committee, and all the NATO military commands. Planning for this exercise began a year in advance. It started with a determination of what concepts or procedures should be practiced. Planning Groups would then begin meeting to produce scenarios that would force the exercise participants to demonstrate their familiarity with NATO operational plans and procedures. Once the exercise started, the Directing Staff or DISTAFF would interject events into the exercise to force the participants to practice the procedures that had been determined ahead of time.

For exercise purposes, I was triple hatted as an exercise planner, Crisis Management Cell leader, and member of the DISTAFF. We developed the timeline and events that would be used in the exercise and the development and submission of recommended actions to respond to the exercise events. As a member of the DISTAFF, we identified what we wanted to do with the ROE, the NATO Precautionary System, or Potential Military Response Options in Crisis. Then we developed the interjects that would be used during the exercise to generate discussion or actions involving these systems. The Crisis Management Cell would receive these interjects, and as the Cell Leader, I would make sure the folks understood how to generate things like an ROE Request, ROE Authorization, ROE Implementation, or activation of a NATO Alert/Precautionary Measure. Then as a member of the DISTAFF, I could influence the entire process if we were not achieving the desired results by injecting certain events at the highest exercise level to try to force the exercise players into the desired situation that would require them to consider certain Alert/Precautionary measures or Rules of Engagement for use in the exercise. They got to learn what the measures entailed and how to execute them. Given the time difference between Europe and the U.S., I spent most of the exercise in SACLANT Headquarters.

One of the more enjoyable aspects of the job was the travel and being introduced to unfamiliar places in Europe. I got to travel regularly to Supreme Headquarters Allied Powers Europe (SHAPE) in Mons, Belgium. I would stay in London and take the train out to CINCHAN HQ at Northwood in the U.K. I spent a week in Oerius in Portugal, which was just outside of Lisbon at IBERLANT. There were regular quarterly trips to SHAPE in Mons, Belgium. In all, I spent about 25% of my time on the road with various conferences and exercises.

The best thing about duty at SACLANT was the exposure to the international community. You were dealing with folks from all different services and countries. The U.S. Navy's philosophy at the time was not to send their best officers to these billets. The Allies, on the other hand, sent their best. For them, it was a chance to broaden their perspectives, and experiences, and get to deal with larger formations and combined forces than they would ever be able to do on their national level. Two of the officers I worked closely with during my time in NATO eventually would become the Chief of Defense (CHOD) in their home countries of Portugal and Denmark, equivalent to the Chairman of the Joint Chiefs of Staff. One of them went on from being the CHOD to becoming the Chairman of the NATO Military Committee. The highest-ranking military man in the NATO alliance.

SACLANT was also the first time I had close contact with Flag Officers. I was a regular briefer during NATO exercises such as CMX or another exercise called OPEN ROAD. Because the ROE and the NATO Alert/Precautionary System involved coordination and consensus at the highest level in the Alliance, I found myself briefing SACLANT, Deputy SACLANT, and various Flags within the HQ on a regular basis. You only needed to screw up once on the international level, and you would be done. By the end of my tour, I was extremely comfortable with this role and received personal cards from the Flags involved praising me for my work, as well as mention of that expertise in my FITREPS. The ultimate compliment came from a Captain who said, "I never saw anybody before as good in Crisis Management in NATO as your predecessor, Dave Nimmer. But you were even better than he was at ROE and the Precautionary System."

While I was serving at SACLANT, the U.S. Navy instituted something called a Selective Early Retirement Board or SERB. The purpose of the board was to identify Navy Officers for separation from the service. When the first board came out, I was surprised at some of the officers serving at SACLANT who were picked for retirement. A Captain who had served as a ship CO and in Major Command was told to go home. Everyone was anxious about who would be next. In the middle of this, I was coming into the zone to be considered for promotion to Captain or O-6. My FITREPS at SACLANT had been outstanding, I was consistently the #1 Commander in the command, but it was NATO, and I was not sure how well that would play in the Navy. Once again, it was time to start the negotiating process for my next set of orders, and for the first time, I would become familiar with the phrase "needs of the Navy."

At some point, the detailer said he thought I would make O-6, so we started talking about jobs. I did not think I was competitive for Major Command because I had been away from the LAMPS community in NATO for the past three years. I definitely was not thinking about getting back into aviation. The detailer asked if I would be interested in a job at the State Department. The senior Navy representative would be rolling at the same time as I was. It would involve working at a more senior policy level than I had before, which interested me. So, I went up to Washington to interview for the job. I did not do well in the interview. The State Department representative knew that I had been working in NATO and asked me what I thought about the U.S. position on the Balkans. I was critical of the failure of the U.S. to take a more aggressive leadership role in getting the Allies to act and said the Europeans were happy to put a fence around the Balkans and let nature take its course. They did not want to get involved. U.S. leadership would be necessary to solve the situation, and it would involve military action. The State Department representative was polite, but I felt that I would not be in the running for the job.

After I told the detailer about "bombing" the interview, I asked what was next. He said that until the State Department rejected my nomination, he could not nominate me for another billet. A couple of months went by, and still no word from State. As my projected rotation date grew closer, I began to worry that all the good jobs would be filled, and I would get the leftovers. Finally, the detailer called, State had rejected me, and he had a job on the staff of the Chairman of the Joint Chiefs that needed to be filled. He gave me the name and number of the incumbent and told me to talk with him about the job and then call the detailer back. I called, and the guy on the job outlined his responsibilities. There was a lot of action officer types of stuff, potentially some speech writing and just responding to the daily needs of the office. He said that based on my background, I would be a perfect fit. I called the detailer back and said I wanted the job.

The detailer said you have the job. But first, I must nominate you for a NATO job in Naples, Italy, at Allied Forces Southern Europe (AFSOUTH). It is a joint duty critical billet, and you are a JSO with a tour in a joint NATO billet, so I have to nominate you. I broke in and said that I had no interest in #1 going to another NATO job and #2 ever living in Naples. The detailer said not to worry, AFSOUTH will never take you. It is a senior Captain's position as the Operations Chief on the staff, and you would be frocked when you go over there. So, I expect AFSOUTH will reject your nomination for being too junior, and then you can go to the job on the Chairman's Staff. I said, but what if they do not reject me, I do not want to go. He said not to worry. When I told my fellow NATO staff officers at SACLANT that I had been nominated for a job at AFSOUTH, they all said, "it is a done deal. With your experience here at SACLANT, they will take you in a heartbeat!"

One day I was in the office, and the Yeoman called me and said he was transferring a call from a Navy Flag officer from Italy. I picked up, and Rear Admiral Hank Giffin said he was pleased to meet me and that he looked forward to me coming to Italy and working at AFSOUTH. I told the Admiral that I really did not want to go to Naples and had been assured that I was too junior for the job. Also, that I was worried about the implications of back-to-back NATO tours on my career. There was also a concern about my daughters, who were in high school. I had heard that the Department of Defense Dependents Schools were not that great. The Admiral acknowledged that duty in Naples was particularly challenging. He suggested that I contact the Flag Officers in the LAMPS community and get their take on whether or not to go to Naples. However, the Admiral said that AFSOUTH had accepted my nomination and I would be receiving orders to Naples.

I hung up and went about contacting two of the Flag Officers I had known in the LAMPS community, one of whom had been my CO in HSL-34. My former skipper RADM Skip Dirren had served in Naples as the Executive Assistant (EA) to Commander in Chief Allied Forces Southern Europe (CINCSOUTH). He said it was a wonderful place, that the tour had been challenging and rewarding, and he encouraged me to take the job. The other Flag, RADM Kevin Delaney, said if the position at AFSOUTH is what the detailer thinks is best for your career, then you should take the job. Now I knew that I could not depend on Flag intercession to try and get the orders to Naples turned off. I called the detailer back and said I was unhappy with the orders to Naples and wanted to know if any other orders were available. He repeated that the job was in a Joint Duty Critical Billet, required a proven Joint Staff Officer, and that I was the only Navy officer who could fill the billet at that time. He said "the needs of the Navy" required me to take that job. Unspoken was that if I did not accept the job, I would be getting out of the Navy. So, I "volunteered" to go to Naples.

A couple of weeks later, I received a call from a friend of mine that was stationed in Naples. Captain Larry Larson had been ahead of me in the Training Command, but we lived in the same housing complex and were both from Ohio. He filled me in on all the "gouge" or required information to make a move to Naples as painless as possible. He also told me that I would not be going to the Operations job but would be going to Plans and Policy, where I would be the Policy Branch Chief. I asked, what happened to me going to the Operations job? He said there were three Navy Captains coming to Naples at the same time, and the jobs that were open were in Operations, Plans, and Policy, and as the Executive Assistant (EA) to the Chief of Staff. He indicated the Ops job was going to a surface warfare officer because he was the senior Captain and the most experienced. The EA job was going to another aviator from the P-3 community, and Larry felt that the P-3 flag in the area had pushed him for that job. I got what was left.

I immediately got on the phone with my detailer and said that I was not going to the billet he had nominated me for, and that if I was not going to the operations job, my orders to Naples should be canceled. He said there was nothing he could do. The receiving command could do anything they wanted with you once you had received orders. So, I packed my bags and bid the family farewell. They would stay in Norfolk so our oldest daughter Carey could graduate from high school. The family would have the year to prepare for the overseas move to Naples. I would become a geographical bachelor for the first year of my tour.

CHAPTER XV

INTO THE BALKANS

ALLIED FORCES SOUTHERN EUROPE (AFSOUTH)

JULY 1994 – APRIL 1997

I left claw marks all the way from Norfolk to Naples. Ironically, being assigned there was the best thing that ever happened to me, although it did not start that way.

I had been to Naples when I was deployed with the Bounty Hunters of HSL-37 Detachment 4, as well as when Linda and I were on vacation. I had also spent a couple of weeks at Naval Air Station Sigonella in Sicily. So, I had some idea of the culture shock I would be experiencing in moving to Naples. But nothing can really prepare you for the chaos the first time you experience it. Nevertheless, I would learn to love the climate, the food, and the people. The noise, the pollution, the grit, the crime, not so much. Naples is like a beautiful woman with dirty feet!

The first order of business was to go to Naples Area Orientation Class. I would complete the administrative check-in process, get briefings on all the services offered by the base, go on an intercultural trip to downtown Naples, and settle in. I was staying at a hotel in Pinetamare until I could find an apartment. It was August, and I quickly realized that when people say that Europe closes in August, it is no joke. But I was staying in a tourist area near the beach, so the vacationers would be up to all hours in the morning, laughing, partying, and riding their Vespa Motor Bikes right outside my window until four in the morning. I had adapted to it when September rolled around. I remember there were a series of thunderstorms on the 31st of August. When I woke up the next morning, the temperature had dropped, and all the people seemed to have disappeared.

Eventually, I found an apartment in a Parco in Monterusciello, outside of Naples. I was the only American living in the apartment building, and the landlady lived next store. She and her husband spoke pretty good English and would use their son or daughter as a backup when they were uncomfortable with their English in talking with me. I had acquired a book on Italian and was trying to get the basics down. The two years of Latin in high school allowed me to translate what was written, but folks spoke so fast that I had a tough time understanding them. "Parlare piu lentamente, per favore!" was the Italian phrase I used most often, "Speak slowly, please!" Even at my best, I spoke Italian in infinitives because I could not conjugate the verbs. The landlady would often invite me to family dinner on Sunday, and I could get about 35% of what they were talking about on a good day. Like most Italians, the whole family smoked. I would purchase cigarettes as a gift for them because they were cheaper at the NATO base in Bagnoli.

My first traffic accident happened about two weeks after I arrived. My sponsor had loaned me her second car, a "beater" that had been passed from family to family for several years. I was looking for a road and was nervous because there were not a lot of street signs in Italy. Instructions used landmarks to help you find someplace. It was not unusual to be told to drive three miles and turn left when you see the "Campfire Girls" by the side of the road. Or turn right when you pass the burning trash bin. Anyway, I did not see my landmark until it was too late and tried to make a left turn in front of oncoming traffic. I misjudged the distance, and the oncoming driver hit me.

We both got out of our cars, and he immediately began yelling and gesturing in Italian. I apologized as best I could in my basic Italian (mi dispiace), I am sorry! Of course, several other cars stopped, and everyone got out, yelling and gesturing in Italian. After about five minutes, the other driver went across the road, bought some water and soft drinks, and returned. Someone in the crowd could speak English, so they functioned as a translator. I told them I was responsible and would pay for the damages. After about five minutes of animated conversation in Italian, the translator said to me, "you must come back here tomorrow at exactly the same time, and we will have this accident." I said, what are you talking about? We have already had this accident. The translator said the other driver was uninsured, which was against the law. So, the other driver would go home, get some insurance, and then tomorrow, we would have the accident after his insurance was in effect. And that is what we did. That was the first time the following phrase popped into my head, "nothing in Italy is easy, but everything is possible!"

It soon became apparent to me that the major decision I would have to make every day was what time am I going to the gym. As the Policy Branch Chief, I had six officers working for me in the air, land, and maritime sections. The Navy Branch had a Greek Navy Captain and a Turkish Lieutenant Commander, the Air Section had an Italian Colonel and U.S. Air Force Lieutenant Colonel, and the Army Section was comprised of a Greek Colonel and a Turkish Lieutenant Colonel. We were responsible for reviewing and updating the Force Plans for AFSOUTH. Under the Force Planning System, NATO staff officers would identify the number and types of forces necessary to defend the Alliance. We would then consult with the nations in the Southern Region to see what they would provide and negotiate with them to fill the gaps. Of course, they would also ensure that their individual national equities were preserved.

I had heard a lot about the relationships between the Turks and the Greeks before I arrived at AFSOUTH. It was surprising how cordial and friendly they were with one another on a personal level. However, when national interests were raised, they had to ensure they made their national position known and would not compromise on it. I had always been able to separate in my mind what was best for NATO as an alliance from what was best for the United States as a country. If I criticized or said I did not agree with the U.S. position, I am not going to be recalled and reduced in rank. For the countries in the southern region, failure to support the national position could be a matter of life or death.

Having settled into the rhythm of the job, headquarters, and Naples, I was surprised when my boss, an Army Brigadier General, called me to his office in the late fall. He shook my hand and said he had a job for me. He wanted me to coordinate AFSOUTH's participation in a Computer Assisted Exercise (CAX-95). The CAX would prepare AFSOUTH and the Allied Command Europe Rapid Reaction Corps (ARRC) for potential operations in the Balkans. The exercise would be used to validate OPLAN 40104 DETERMINED EFFORT, which would provide NATO support for the United Nations' withdrawal from Bosnia Hercegovina. 41 Flag Officers and over 1,000 personnel from AFSOUTH's Principal Subordinate Commanders would participate. My job was to ensure that the training was coordinated with the various departments at the respective headquarters and that the exercise goals and objectives were achieved. My stint at SACLANT had prepared me for the job. We would conduct the exercise at the Warrior Preparation Center, located at Einsiedlerhof, Germany. After we had done all the exercise planning and arranged for the staff to go to Germany, I was told that I would be working as the Chief of Staff for the Directing Staff or DISTAFF. I would get to torture my fellow AFSOUTH and ARRC staff officers by running the exercise.

A major blowup came when the AFSOUTH Chief of Staff (COS) Lieutenant General Marvin Covault, U.S. Army, learned that we would not be exercising continuously for the entire twenty-four hours every day. His objective was for the staff's battle rhythm to consist of continuous exercise play with minimal time for sleep to simulate what would occur in combat conditions. The exercise had been set up so that at the end of the normal workday, the exercise would be paused; the players would get something to eat, get a good night's rest, and then resume the exercise play where they left off the next morning. This was the normal battle rhythm for a NATO exercise. The COS was extremely unhappy when he heard this and wanted to know why I had not ensured continuous play as he had previously directed. If my boss had told me that 24-hour play was required, I had forgotten it. So, I told the COS that I could not remember being told that we had to go for twenty-four hours. For a few minutes, it looked like I would be fired and sent back to Naples. While the COS was wire-brushing me for being such an incompetent staff officer, the U.S. Navy Rear Admiral in charge of the DISTAFF stepped in and assured the COS that we would make sure the staff was suitably stressed during our normal workday. He took the heat off of me. The COS had a reputation for being very exacting in what he expected from his staff and did not hesitate to express himself when he felt his expectations were not being met. I figured I would hear about it again when we got back to Naples.

The exercise itself went off well and served as an opportunity for the respective staff to get to know one another. Personal relationships were developed among the senior flag officers and their staffs that would prove invaluable when NATO forces were called upon to execute the Dayton Peace Agreement in Bosnia Hercegovina. I spent a lot of time with the U.S. Army folks from our Planning shop and developed a better appreciation for the staff work they would crank out. It was the first time I had worked with the "Iron Majors," who formed the backbone of any Army

Staff and experienced how dedicated and hardworking they were. They could develop a synchronization matrix or a phased operations plan at the drop of a hat. They would be doing a lot of synchronization and planning over the course of the next year, but we did not realize it at the time.

Following our return from Germany, I submitted folks for awards, put together some lessons learned, and settled back into being the Policy Branch Chief. We were working with several nations in the region as well as NATO's Naval Striking and Support Forces Southern Europe (STRIKFORSOUTH) on producing a Combined Amphibious Forces Mediterranean (CAFMED) and Mine Countermeasure Forces Mediterranean (MCMFORMED) policy documents. There were also meetings with the nations to go over their force proposals and how they would affect force planning.

In March, my boss in Plans and Policy asked me to come to his office again. When I walked into the office, he came around from his desk and shook my hand. I joked that the last time he did that, I wound up heading the AFSOUTH participation in the CAX-95 exercise at Einsiedlerhoff, Germany. He laughed and said it was nothing that drastic. Then he said I would become the Executive Assistant (EA) to the Chief of Staff (COS) at AFSOUTH. I am sure I looked like someone had hit me with a two-by-four. I asked, what happened to the current E.A.? The General said that the current E.A. was being relieved and would go into operations. I was to wrap things up in Policy and get over to the COS's office as soon as possible. I reminded the general that the COS and I had some issues up in Germany during the CAX-95 exercise, and I was not sure that he would really want me as his new E.A., especially as I would be the third officer to hold that job following an Army Colonel and a Navy Captain. The General told me the COS had asked for me specifically. I did not know if I were a placeholder until some Army

Colonel could arrive or if I was to be the permanent E.A. I would find that out when I got into the office. My relief with the incumbent Navy Captain EA was short. He was told to get down to the operations division in the tunnel as soon as possible. I would learn about being an E.A. and being an E.A. in a Joint (Army, Navy, Marines, Air Force) and Combined (Sixteen NATO nations) environment through on-the-job training.

The COS, LTG Marvin Covault, provided me with a terrific opportunity to gain experience about being an E.A. I did everything he told me to do when he told me to do it and exactly the way he told me to do it. We got along fine during my brief time with him. He knew what he wanted, did not vacillate or hesitate, and so all I had to do was execute. I learned a lot about how the staff worked. Who were the major players? Who delivered, and who did not? Which ones to trust, and those who had to be watched. I was also learning about how important the personal relationship was between the COS and the Commander In Chief (CINC), as well as the Assistant Chiefs Of Staff.

The first and most important lesson I learned was that I was not the COS. I did not wear his three stars. I did not speak for him. My job was not to guess what he wanted but to ask him what he wanted, make sure I completely understood what he wanted, and then transmit that direction to the staff. When the staff indicated they were not sure what the COS wanted, it was my job to go in and ask for clarification or additional guidance. I was also a shock absorber between the COS and the staff. When the COS was having a difficult day, it was my job to ensure he took it out on me and not the staff. I had to be the honest broker. My word was my bond, and I could not lie or mislead. If I did not know, I said so and then went to the COS to find out the answer. Over time, and with experience, I could offer what the COS had said or indicated he wanted to be done in each situation, but it always came back to, "I'll go ask the

COS and get back to you with an answer, or, let me put him on the phone, and he'll tell you exactly what he wants."

My working routine changed from when I go to the gym to, will I ever get to go again. LTG Covault liked to get to work early. He would arrive in the operations spaces at about six in the morning, then come into the office, attend the morning meeting with the ACOS and the CINC, and then proceed with a series of meetings, briefings, and phone calls. He did not like to stay late, so he would take work home with him. When I would arrive in the morning, the General's driver would come to the office after he dropped the COS at operations and give me all the stuff that the COS had taken home the night before. I would go through it and assign actions and taskers based on the COS's notes on the individual documents. The next day we would rinse and repeat. I grew increasingly comfortable working for the COS and respected him tremendously. I got to the point where I would tease him about things. Just part of the Irish Catholic fatalism, sticking your head into the lion's mouth. I took a cartoon from The Far Side by Gary Larson. In the cartoon, the first caption shows a dog and its' Master. The Master is talking to the dog, and you see what he is telling the dog to do. The caption was, "what the Master says." In the second caption, you see the dog listening, and the caption says, "what the dog hears," which is blah, blah, blah, blah, blah. I took the cartoon, substituted what the COS says and what the E.A. hears, and slipped it into his stack of papers to read. About half an hour later, from inside the COS's office, I heard, "very funny."

The physical setup of the COS's spaces had me sitting right outside his office. I was literally the gatekeeper. On the other side, his room opened into the CINC Admiral Leighton "Snuffy" Smith's office. They would talk with each other throughout the day. I soon learned that they had a good personal relationship as well as a strong professional one.

I also figured out quickly that one of my main objectives was to have a good working relationship with the CINC's E.A., Captain Rusty Petrea. Again, I listened a lot, asked questions, embarked on a vertical learning curve, and did exactly what Rusty told me to do. I learned that the E.A.'s often were used to pass on information to one another for their respective bosses. That way, issues could be raised and discussed without having the principals involved. In my case, it was telling the COS that the CINC's E.A. said the following. Then I would go back to Rusty and let him know how the COS felt about the issue.

The golden rule among E.A.'s was "no surprises." Our bosses needed to know exactly what was going to happen, the pros and cons, who was on what side, the possible outcomes, and the recommended course of action. The worst thing in the world was to get "the look" during a meeting when you knew that you had not anticipated and prepared the COS for the meeting properly. Whether or not it was your fault, you took the blame and promised to do better the next time. After about three months, LTG Covault rotated back to the States to retire.

The new COS, LTG William G. Carter III, was coming from command of the 1st Armored Division "Old Ironsides" in Germany. My task was to ease his transition into living in Naples and learning the ropes at AFSOUTH. LTG Carter had been an Aide and the Executive Officer to the Chief of Staff of the Army, so he knew my job inside and out and how a COS should run a large headquarters. Early on, I asked him if he wanted me to remain as his E.A., and he said yes. My first big decision was who to recommend as his Aide. The previous Aide was moving on, so I put together a list of candidates. He interviewed them all and then asked me for my recommendation. I knew my choice would cause the most heartburn because Major Jim Moentmann would be moving from my old Plans and Policy shop. Jim was an outstanding officer and would leave a big hole in their organization. He had been my assistant during the CAX-95 exercise, and as an Army Aviator, we worked together well. I told the COS who I recommended and that the ACOS for Plans and Policy would not be happy. The COS went with my recommendation, and our team was now in place.

EA TO THE CHIEF OF STAFF

Being an E.A. means providing your boss with the best possible working environment. You must learn how the COS wants to operate. Does he prefer to get his information from reading, briefings, meetings, etc.? The new COS had a different operational mode than the previous one. He liked to get to work a little later in the morning and then go to the CINC's daily briefing. He would stay in the office until the last meeting took place, the last phone call was made, all the briefings were completed, and all the documents had been reviewed. Whenever all that had been completed, we would go home. The time varied from day to day, but in general, it was usually between 1800 and 1900, following an 0700 start. It took Jim Moentmann and me about a month to figure the work schedule out. At about 1700, we would take the documents out of the COS's inbox, put them in the safe, and lock them. After about three days, LTG Carter walked out one afternoon, looked at me, and said, "where are you hiding the documents? They are in the safe, aren't they?" I nodded, and he smiled and went back into his office. But we did start going home a little earlier than we had been doing.

I had to adapt my way of doing business to provide him with the best possible support. Because there was a minimum time between the COS's arrival and the CINC's briefing, I would put together a newscast of the information he had to be familiar with before the day started. I would go over what had happened overnight operationally and what was on the day's schedule, highlight the important meetings, give him a synopsis of the documents in his inbox, and recommend the ones he should read prior to the meeting. Then I would close the door and give him time to digest all of that prior to the meeting with the CINC.

With the previous COS, if someone arrived and wanted to meet with him, I would just pop my head into the office and tell him who was there and what they wanted to discuss. He would indicate whether or not he wanted to meet with them at that time. We used to refer to these impromptu meetings as drive-bys. LTG Carter got a daily card that outlined his meetings for the day. One day early on, someone wanted to meet with him but was not scheduled. I walked in and told him who wanted to come over and what they wanted to talk about. He pulled out his card, looked at it, and said, "he's not on the schedule." I waited for further clarification, but he went back to working on his papers. I went back to the outer office and walked over to Jim Moentmann's desk, and explained the situation to him. I did not know what to do. Jim said, I got this and went into the other office where the administrative specialists, SFC Siders, and SGT Oglesby worked. He returned a few minutes later and went into the COS's office. Jim came back out and said the COS would meet with the individual. I asked him, what did you say to the COS? Jim replied, nothing. I just took him a revised schedule card with the new proposed meeting on it, and he said OK. If it was on the card, we did it, but the card had to contain the latest information.

I went to the Navy Exchange and bought the longest telephone receiver cord I could find and installed it on my phone. It allowed me to walk into the COS's office while I was talking with someone on the phone and allowed him to be able to hear my part of the conversation. He could give me direction on what he wanted to be done without necessarily talking to the individual. An emphatic head shake or thumbs up gave me guidance on how to respond to whoever was on the line. Sometimes he would get on the phone in the middle of the conversation. One day I got a call from an ACOS who was seeking guidance on a particular issue. I took the phone into the COS's office and outlined the issue and the practical solutions. The COS indicated which one he favored with hand signals. I told the ACOS that the COS desired that he does such and such. The ACOS began to argue with me about what I had said. The COS picked up his phone and told the ACOS that I spoke for him and that when I directed something to be done in his name, it was because he, the Chief of Staff, had previously directed me to do so. It did not take long for the folks at the H.Q. to figure out that I was not freelancing and that I usually was standing in the COS's office when they were talking to me.

Against this backdrop, the situation in the Balkans continued to deteriorate, and NATO's involvement increased.

Some historical background. On the 8th of June 1993, NATO's North Atlantic Council and the Council of the Western European Union approved a combined concept of maritime operations in support of the United Nations arms embargo against all republics of the former Yugoslavia and sanctions against the Federal Republic of Yugoslavia (Serbia and Montenegro). Operation SHARP GUARD was undertaken in support of the United Nations Security Council Resolution 713. Operational Control of a combined NATO /WEU Task Force was delegated through the Supreme Allied Commander Europe (SACEUR) to the Commander Allied Forces Southern Europe (CINCSOUTH), who delegated Operational Control to Commander Naval Forces Southern Europe (COMNAVSOUTH) located in Nisida just outside of Naples.

NATO Air Operations were phased in over a period of three years. Initial NATO involvement was in the establishment and monitoring of a no-fly zone over Bosnia-Hercegovina in support of U.N. Security Council Resolution 781 in October 1992. In March of 1993, U.N. Security Resolution 816 authorized enforcement of the no-fly zone over Bosnia-Hercegovina. Only fixed wing and rotary wings flights authorized by the U.N. were allowed. On the 12th of April 1993, NATO began operation DENY FLIGHT utilizing fighter and reconnaissance aircraft based in Italy and aboard aircraft carriers in the Adriatic. NATO's first military engagement occurred on the 28th of February 1994 when four Serbian aircraft that were violating the no-fly zone were shot down. NATO forces began providing Close Air Support to Safe Zones and United Nations troops.

Operation DELIBERATE FORCE took place between the 30th of August and the 20th of September 1995. The air strikes were concentrated on reducing the combat effectiveness of Serbian forces, increasing their difficulty in effective command and control, and evening the balance of power in the Balkans. The ability of the Serbs to continue to prevail over the Croats and Bosnians was called into doubt.

The warring parties eventually got together in December 1995 and signed the Dayton Peace Accords. NATO forces were charged with the implementation of the military annexes of the Peace Accords, known as the The General Framework Agreement for Peace (GFAP), in Bosnia Hercegovina. NATO forces entered the former Yugoslavia on the 20th of December 1995 as the Implementation Force (IFOR). IFOR's headquarters were initially established on the grounds of the United Nations Protection Force (UNPROFOR) in Zagreb, Croatia, on the 18th of December 1995. The H.Q. moved to The Residency in Sarajevo, Bosnia Hercegovina, right after Christmas. NATO forces initially focused on the separation of the warring factions along the lines agreed in the DPA, enforcement of a cease-fire, and establishment of weapons collection areas.

During these events in the Balkans, the headquarters was focused on learning as much as possible about what our responsibilities would be under the peace agreement and what our roles would be in enforcing it. One of the members of the AFSOUTH staff was in Dayton and provided us with a back channel about what was going on. We could provide feedback to him on the proposed military actions and whether or not they made operational sense. This information was then passed on to the planners in AFSOUTH and the ACE Rapid Reaction Corps (ARRC) Headquarters so that they could begin formulating plans to provide for the movement of their headquarters and NATO forces into the Balkans. We were also busy trying to produce what the headquarters staff would look like. What functions would go forward, and which would remain behind? Our initial planning was based upon the Headquarters remaining in Naples and providing guidance to the Commander of the ARRC on the ground in the Balkans. All the command, control, and communications were set up in Naples, and we did not want to have to replicate that capability in theater. However, there was a great deal of pressure from SHAPE and NATO HQ to have the CINC in theater. The initial compromise was that IFOR HQ would move to Zagreb and occupy the former headquarters of the United Nations Protection Force (UNPROFOR), providing direction to the NATO Ground forces from there. However, it was not long before the decision was made to move once again, this time to Sarajevo and The Residency.

The staffing of the IFOR headquarters was based upon the number of racks (the navy term for beds) that were available at the time as well as what functions would be performed better by remaining in Naples. The idea was to move the minimum number of staff officers forward as necessary. Eventually, we settled on the following going forward: intelligence, operations, computer information systems, legal, public affairs, and civil-military cooperation. Logistics would be assumed by COMMZ in Zagreb. The Plans and Policy shop would remain in Naples, providing daily updates on planning and synchronization with individuals coming to Sarajevo on an as-needed basis for short periods of time. Budget and Finance would be split between Naples and Zagreb.

We arrived in Zagreb and moved into the Headquarters of UNPROFOR. It was surreal. We were in our battle rattle, helmets, vests, weapons, etc., staying at the Intercontinental Hotel in the middle of a major metropolitan city that had not been significantly affected by the war. Zagreb had been struck by rockets earlier in the war, so there was the element of a possible attack.

My routine remained the same. Stop by the Join Operations Center (JOC) and get briefed on overnight events, scan the message traffic, consult the daily schedule, and draft my briefing for the COS. One thing that was different is that I started screening press reports from throughout the Balkans. These newspaper and wire service reports provided background on current events and trends within the area of operations. A wire service might report an incident in Banja Luka or Mostar. I would cross-check with the operations reports to see if there was any correlation. Often there was no mention of the incident in the intel or operations reports. If the incident seemed serious, I would point it out to the COS and let him decide whether to bring it up with the CINC. The Intel folks were great at telling you what happened last week. Not so good about what happened yesterday or last night. Unfortunately, those were the incidents the press would be interested in, and the CINC or COS would be asked about. They would look foolish if this was the first time they heard about something happening in theater. After a couple of weeks, the intel action officers came to me and asked, where is the COS getting this stuff about what is going on in theater? I told them. I provide the information to him based upon reviewing the media reports from within the theater. I suggested they do the same for their briefers so that when the CINC or the COS would ask about an incident in a briefing, they could respond with what they knew, that they were working on the issue, and specify when they would have a report for leadership.

It was soon apparent that the H.Q. would be moving again, this time to Sarajevo. My job would be to remain in Zagreb for a couple of days as the communications link and to provide continuity until Sarajevo was fully online. After a few days, I reported to the Zagreb airport and hopped on a flight to Sarajevo. There was the corkscrewing approach into the airport. Nothing that a LAMPS helicopter pilot had not experienced before. Sarajevo had been severely damaged during the war. As an army officer said to me, there was no shortage of ammunition in this war. It seemed like every building had been damaged.

The ride from the airport to the H.Q. at the Residency was quick, and there were no other vehicles on the road. The Residency had been a guest house for senior members of the Yugoslav Communist Party. It was located between the U.S. Embassy and a barracks that was used by the Bosnian Intelligence Service and a guard force. I arrived and went to our offices. They were trailers that had been stacked on top of one another in the courtyard. We had three assigned to us. LTG Carter had his own space, Jim Moentmann and I shared the next trailer, and the last one was for the Admin Sergeants and security.

EA & AIDE TRAILER OFFICE IFOR HQ SARAJEVO

The COS lived in the Residency with the CINC, who had his offices and staff with him in that building. We lived on the top two floors of a four-story barracks that had been damaged during the war and had formerly housed Bosnian Military Intelligence. There were three officers to a room. There was no central heating and sandbags where the windows used to be. You had to walk across the compound to take a shower, shave, or go to the bathroom. We took our meals in the Residency. For an Army guy, this was easy. For a Navy guy, this was a rude awakening. Our days were punctuated by occasional bursts of gunfire, which LTG Carter and Major Moentmann would classify as heavy or small arms, guess what caliber, and how far away from our present position. This was all to assure the Navy Captain Helicopter Pilot, who was hiding under his desk, that we had nothing to worry about.

We also enjoyed being able to have an occasional alcoholic beverage if we so desired. The U.S. European Command had instituted General Order #1, which said no U.S. troops would consume alcohol in Bosnia. Admiral Smith, as the Commander of IFOR, took the position that we were NATO troops under NATO command, and General Order #1 did not apply to the NATO IFOR Headquarters. We had alcohol in the office, and the COS would make it a point to offer a drink to any U.S. personnel who came to visit the headquarters. They would always decline. After LTG Carter left for the day, we would light the drinking lamp after business hours were over. We stayed in the office because the COS had access to the American Forces Network, and we could watch American television programs and sports. We'd vote on what to watch, open a couple of cold ones, or in my case, mix a Rum and Coke, and settle in for the evening. If there was nothing on TV, I would go back to my room and read. The day started at 0600 and ended at 1900, seven days a week unless the COS had something else on his schedule.

My roommates for most of the stay in Sarajevo were Air Force Colonel Steve Bryant, who was the AIRSOUTH representative working in our operations center, and Army Colonel Jim Burger, who was the AFSOUTH Legal Advisor. When I would get back to the room, we would go over how our respective days had gone. Their observations provided me with valuable context as to the day's events and how our office and the Headquarters, in General, were perceived by folks outside our compound. Jim Berger worked closely with the IFOR Civil Military Affairs folks and the Non-Governmental Organizations, so he spent most of his days outside of the headquarters and had valuable insights as to how IFOR was being perceived by folks in the country. Steve and Jim would not hesitate to bring up issues regarding the headquarters and how we were doing. They knew that if they had issues and brought them up, I would take them to the COS and provide feedback on what his position or recommendations were. All information was good. I might not like it, but I needed to be aware of what was going on and what was being said. I also knew that if you shot the messenger, they would never provide you information again, or even worse, would just tell you what they thought you wanted to hear.

LTG Carter spent most of his day trying to get all the various factions and organizations in Bosnia to play well with each other. He developed exceptionally good relationships with the U.S. Ambassador to Bosnia Hercegovina, John Menzies, and Jock Covey, the Chief of Staff to the United Nations High Representative for Bosnia and Hercegovina, Carl Bildt. There was a side gate between the Residency and the U.S. Embassy that we referred to as the "Carter Gate" that we used to take when we had meetings in the embassy. The COS would have meetings with various other organizations and occasionally would go out to dinner or attend social functions in Sarajevo. This required a day reconnaissance by the Aide and the driver to check out the sites. The COS would attend with the driver, security, and depending on the event, the aide or myself.

There was an inevitable crisis over compliance by the parties with the provisions contained in the Dayton Peace

Agreement. Our focus initially was to separate the factions, identify the weapons collection areas, and provide security. The decision to go into Bosnia in the winter benefited IFOR. The feeling was that given the harsh winter conditions, the factions had enough on their hands with just staying warm. That gave IFOR time to move in and establish control.

We were still responsible for running AFSOUTH simultaneously while directing IFOR. That entailed video teleconferences, phone calls, and meetings to deal solely with issues affecting the southern region of NATO. The net effect was that there was never a day off. Initially, we spent most of our time in Sarajevo. Over time we began to split the week between Sarajevo and Naples, but we were still working seven days a week. When we were in Naples, and it was the weekend, we used to wear civilian clothes at work to remind ourselves what day of the week it was. We also tried to work a half day on the weekends in Naples if it was possible.

Another challenge was meeting the unique command structure that had been set up within IFOR. The Commander IFOR (COMIFOR, Admiral Smith) and the Chief of Staff (COSIFOR, Lieutenant General Carter) were U.S. officers serving in NATO billets. The Deputy Commander in Chief IFOR (DCOMIFOR, General Janvier) and Deputy Chief of Staff (DCOSIFOR, MGEN Fleury) were French and had no NATO jobs. The command structure provided a way for the French military to monitor what was going on in the theater and to raise any objections to policies through political channels at the NATO HQ in Brussels. I believe that one of the reasons the COS spent most of the time in Sarajevo was that General Joulwan, who was dual-hatted as the head of the U.S. European Command (EUCOM), and as the NATO Supreme Allied Commander, Europe (SACEUR) felt more comfortable having an Army Flag Officer in theater that he could check in with when COMIFOR was not in town. The COS was able to supply information and assurance to General Joulwan on how things were going in Bosnia based on his previous combat experience. LTG Carter had been an Assistant Division Commander with the 1st Infantry Division during the first Gulf War, as well as having served in Vietnam.

COSIFOR and DCOSIFOR were never collocated in the same place at the same time. When one was in Sarajevo, the other was in Naples. Both used the same office in Sarajevo and Naples. As the EA, I would coordinate between the Generals and the offices and try to sort out any issues. It was my first time dealing with the French Military on a daily basis, and I was impressed. We knew the French were there to look over our shoulders and report back to France on what was going on. They knew that we knew that was what they were doing. Over my time in NATO, I grew to appreciate the difficulties of reconciling my NATO hat with my U.S. hat. You can tell in the body language and the actual words used when you are getting a "national" position, i.e., this is my nation's position on this issue, and I do not agree with it, but I do not have any choice. Anyone who has served in NATO has been in that position.

We had one U.S. officer or senior enlisted sergeant in the office in Sarajevo when the DCOSIFOR was there to provide continuity and to sort out any issues that would come up. Back in Naples, the Special Assistants to the Chief of Staff (SPACOS) would provide support to the French when the COS and the rest of the office was in Sarajevo. I tried to rotate the staff from the COS office to Sarajevo whenever possible. There were several reasons: one, it provided them with a tremendous professional opportunity to see and learn about the NATO Alliance as it engaged in its first out-of-NATO area operations; two, it provided a break for Jim Moentmann and me so that we could take some time off; three, they got a tax break on their payments because they were serving in a combat zone. They really did a fantastic job of keeping things going in Naples while we were in Sarajevo. They also were really good at sorting out the things that really required the COS's attention and what could be managed at a lower level. I had peace of mind because I knew they could help with any family emergencies that might arise. It was Naples. What could go wrong?

COS IFOR STAFF AT HQ IN SARAJEVO

L-R PRIVATE DOENITZ, SFC SIDERS, LTC MOENTMAN, SSGT OGLESBY, LTG CARTER, CAPTAIN COYNE

Lurking in the background as the months went by was the question, "how long are we going to be here?" The President of the United States indicated that U.S. forces' participation in IFOR would only last for a year. We were not in Bosnia long before it became apparent that the mission would require more than a year. How would the U.S. reconcile the problem? We will change the name of the force and give the force a new commander. So, Implementation Forces (IFOR) would become Stabilization Forces (SFOR), and the U.S. Army would take over command.

There was always an undercurrent of discontent and dissatisfaction with the idea that operations in Bosnia were commanded by an Admiral and not a General. These were ground operations that could potentially result in combat. Shouldn't the officer in command be a soldier? Although Admiral Smith and later Admiral Joe Lopez were the Commander, the actual command and control of the land operations was in the hands of Commander Allied Command Europe Rapid Reaction Force or COMARRC, Lieutenant General Sir Michael Walker, a United Kingdom Army Three Star. COMIFOR's job was to provide overall guidance based upon the direction provided by SACEUR, who received his marching orders from the NATO Military Committee and the North Atlantic Council sitting in Brussels.

My impression was that COMARRC was happy working for a Navy Admiral, especially Admiral Leighton "Snuffy" Smith, later Admiral Sir Leighton Smith. As a Naval Aviator, the Admiral's style was to provide general guidance, ask specific questions when necessary to assure himself that the guidance was understood and would be executed, and then let the Commander do his job. Admiral Smith was not inclined to get into COMARRC's business on an operational or tactical level. That would not have been the case if COMIFOR had been an Army Officer. Also, LTG Carter provided a conduit between Admirals Smith/Lopez and Lieutenant General Sir Michael Walker, COMARRC. LTG Carter provided a sounding board for Admirals Smith/Lopez on these operations and provided valuable feedback to COMARRC from a land combat veteran's perspective.

Operations in Bosnia raised issues of "national caveats" for the nations involved in the operations and differences in services cultures in the way they executed those operations. A nation committed forces to NATO operations but reserved the right to determine which kinds of operations those forces would engage in.

In Bosnia, that could involve crowd control and the use of tear gas to control a crowd. Certain nations would forbid their forces from being used in these kinds of operations because they considered tear gas to be a chemical weapon. The U.S. would regularly criticize nations for having these caveats, but the U.S. also had them.

U.S. forces would only operate in certain areas. U.S. forces were perceived as being more responsive to orders received from their national command, U.S. Army Europe, than by COMARRC. U.S. forces were heavily involved in force protection and deployed their forces accordingly. U.S. forces always wore full "battle rattle," helmets, vests, weapons, etc. That applied to all ranks. By contrast, senior U.K. commanders made it a point not to meet with the leaders of the former warring factions in full combat gear. They felt it sent a signal to their adversaries that they were not afraid of them and that if something happened to them, another U.K. officer just like them would take their place. Admiral Smith made it a point not to wear his combat gear when meeting with the senior members of the former warring factions.

The heavy U.S. emphasis on force protection was highlighted when they ran out of forces to provide surveillance on some sites in their area of operations that COMIFOR wanted to add to the list. Admiral Smith was puzzled and asked why there were not enough troops to provide the surveillance. We learned that U.S. policy was to provide a 4-1 ratio in their operations. If you needed one soldier to take control of a checkpoint, four would be assigned. If you needed a team to provide overwatch or surveillance, four teams were assigned. With those provisions, it became obvious why the U.S. had run out of assets to provide surveillance. No one on the U.S. side had provided that information to IFOR.

Over time the situation in the headquarters began to become more livable. The first two floors of the former barracks that we were sleeping in were converted into office spaces, and we moved from the trailers to the barracks. The COS had his office, there was a larger reception area where the Aide and I worked, and the Sergeants had an office directly across from us. The showers and bathroom were operational in the barracks, so we did not have to go across the compound anymore.

Life settled into a routine that was punctuated by crisis as the former warring factions, Croats, Bosnians, and Serbs sorted their way through compliance with the Dayton Peace Agreement. At any given time, IFOR was being assessed throughout the country as to how stringent our enforcement of the peace agreement would be. One of the crucial issues became the capture of Persons Indicted for War Crimes or PIFWCs. The top two wanted men were General Ratko Mladic and the former Republic of Srpska President Radovan Karadzic. There were reported sightings of these war criminals all over Bosnia, and the media and the politicians wanted them taken into custody. IFOR's mission did not include searching for or arresting PIFWCs. We were authorized to detain them if we came across them in the course of our normal duties. Karadzic and Mladic were not in the habit of identifying themselves while going through IFOR checkpoints. This became a source of friction between the IFOR Commander and the international community. When pressed on the issue, Admiral Smith responded he would go after the PIFWCs as soon as he was ordered to do so by the North Atlantic Council. He did not have the authority to order a hunt for them. We believed that the diplomats in Brussels would be delighted for Admiral Smith to do this on his own, and if it went well, the politicians would claim the credit, but if it went badly, they would condemn him for exceeding his authority.

Admiral Joseph Lopez arrived in July of 1996 to relieve Admiral Smith as the Commander of IFOR. Admiral Lopez was a surface warfare officer, and his style of command was different than

Admiral Smith's. As the new guy on the block, Admiral Lopez would ask a lot of questions about why operations were conducted in a certain way. Folks in the ARRC would complain about backseat driving by COMIFOR and ask why the Admiral required that level of detail on normal operations. We told them the Admiral had to feel out his subordinates and gain confidence in their judgment and recommendations. For me, it meant that I had to learn the operating style of the new COMIFOR EA, Captain Ken Heimgartner, an F-14 Radar Intercept Officer, and establish a new battle rhythm. The process went smoothly, and the transition was invisible to the IFOR Headquarters Staff.

Before you knew it, we were making plans to transition over to SFOR. President Clinton's pledge to have IFOR out of Bosnia in a year would be met. The new COMIFOR, soon to be COMSFOR, would be the U.S. Army commander in Europe, General William W. Crouch. It became apparent that profound change was coming. The size of the SFOR HQ staff would be larger than IFOR's. The information requirements for land operations by the SFOR HQ would be much greater than had been the case in IFOR. The U.S. Army was in charge,

and there would now be a properly run H.Q. with the requisite ability to manage the Multinational Divisions in Bosnia.

No one in our office was sad to leave Bosnia. The departure was uneventful. Our workload was cut in half, and we could now focus on AFSOUTH business and work a normal week. For me, it was time to focus on what would come next in my career.

CHAPTER XVI

MAJOR AVIATION COMMAND

NAVAL SUPPORT ACTIVITY NAPLES, ITALY

MARCH 1997 – JUNE 1999

During September in Sarajevo, I received a phone call one day from the Navy Detailers in D.C. They said congratulations, you screened for Major Aviation Command! Major Aviation Command in the Navy is the last step before you are considered for Admiral. My problem was that I had seen the Major Aviation Command list when it came out, and my name was not on it. The detailer explained that there was an "alternate list" of folks who would be used to fill any gaps that were created in the original Major Aviation Command screen list.

It took a minute or two for all of this to sink in, and I asked the detailer, when do I have to submit a preference list for major command? He said no list would be necessary. I had two choices, Naval Support Facility Diego Garcia in the Indian Ocean or Naval Air Station Guantanamo Bay, Cuba. There was another pause while I thought about my choices. Is there any way I could postpone the selection until next year or wait to see if any other billets opened up? The answer was no. Those were the options, take em' or leave em.'

While the prospect of Major Aviation Command was exciting, the choices left me feeling disappointed. This was the pinnacle of my Navy career, I would be a base commander, but the personal cost would be tremendous. I had gone to Naples and

AFSOUTH and left the family behind in Virginia Beach so that my eldest daughter Carey could complete high school. So, I had spent a year unaccompanied in Naples before the family joined me. They arrived in the summer, and I left for Bosnia and IFOR in December. We had been separated for the better part of two years, and Diego Garcia was on a one-year unaccompanied tour. Guantanamo Bay was different; I could bring the family. But the kid's education would take a hit. The schools in Naples were not the greatest, and Gitmo would not be much better. The detailer said I had a couple of days to decide on what I wanted to do.

Fortunately, we would be going back to Naples for a few days, and I would be able to talk all this over with Linda. It was not what we had expected in terms of geographic locations. It was not happening at a particularly stable time in our family's life with one daughter in college in the states and two daughters in Naples American High School, coupled with all the time I had been away and the inherent stresses that places on a family. After we had discussed it for a couple of hours, I asked Linda, what do you want to do? Her reply was the same as countless Navy wives have said over time immemorial, do what you think is right, and we will support you in that decision. She had said that to me before, through the moves to D.C., Hawaii, Newport, back to Hawaii, Norfolk, and now Naples. But there was something in her voice that was different, and I realized we were at a breaking point for our family. I called the detailer back and once again tried to see if I could put off a decision. He said it was Dodge (Diego Garcia) or Gitmo (Guantanamo Bay). I said I would turn down Major Aviation Command. He told me to draft a letter of refusal and send it to the Bureau of Naval Personnel.

From: Captain John J. Coyne, USN

To: Captain Robert E. Besal, Director, CAPT Assignment PERS-43

Subj: MAJOR AVIATION SHORE COMMAND

Ref: (a) MILPERSMAN 1810270

1. Per reference (a), I hereby decline major aviation shore command. My reasons are primarily personal. Professionally, I believe I would have contributed significantly to the Navy by assuming command of either NAS Guantanamo Bay, Cuba, or Diego Garcia. However, personally, I felt accepting these orders would force me to choose between the Navy and my family. A brief bit of background. I arrived at AFSOUTH in August of 1994 and elected to do the first year of my three-year tour unaccompanied to allow for my daughter to complete her senior year in high school. Following my family's arrival in July of 1995, the AFSOUTH headquarters was actively engaged in preparing for deployment to Bosnia Hercegovina, which we subsequently accomplished in December of 1995. Since that time, I have spent about 70% of my time deployed to Sarajevo, with the remainder in Naples. As the Executive Assistant to the Chief of Staff, a three-star Army General, since March of 1995, I have found this tour professionally rewarding but personally taxing. In regard to Diego Garcia, having already done a one-year unaccompanied tour, I did not wish to do so again. As far as Guantanamo Bay is concerned, subjecting my family to three years in Gitmo following two years in Naples was too much to ask. If I was able to defer slating to next year to compete for a stateside assignment, or if this assignment had come following three years in Norfolk, I would have accepted. Therefore, so as not to impose an additional burden on my family, I must reluctantly and regretfully decline the aviation major command.
2. I recognize my declination will be forwarded to the Deputy Chief of Naval Personnel and will become a permanent part of my record.

Very Respectfully,
J. J. Coyne
CAPT USN

Even as I submitted this letter, there was something in the back of my mind that said, "you will get another chance."

After turning down Major Aviation Command, I went back to the business of being the best EA I could be. In the late fall, I began to discuss what my next assignment would be with the detailer. Having turned down major command, I knew the options would not be good. We talked a little bit about being the Commanding Officer of an ROTC Unit, and that sounded ok. The COS and I talked about taking a job at NATO HQ in Brussels. The U.S. Military Representative (MILREP) to the NATO Military Committee was a friend of his and would be visiting Naples, and LTG Carter wanted to know if I would like to have dinner and see if there was a good fit. The dinner went well, and I called the Navy detailer to ask about a Navy Commander job on the MILREP's staff in Brussels. The detailer said I could not have the job. I was an O-6, and the job in Brussels was for an O-5. My response was, "I'm an O-6 who turned down major command and is a dead man walking. This is a personal request from an Army O-9 to be on his staff. We ought to be able to make this happen." The detailer said the request was coming from the wrong color officer. I said, "what?" He clarified that he could not assign an O-6 to an O-5 billet, especially working for an Army Flag Officer. I asked, if the request was coming from a Navy Flag Officer would the response be different? He said it depended on the Flag Officer. I said, so you want me to bring the heat? He said yes. I said would you like a call from a three-star or a four-star? There was a momentary pause on the other end of the phone, and he said, do what you have to do.

I had a great working relationship with the EA to Admiral Lopez and the NATO STRIKEFORSOUTH Chief of Staff to Vice Admiral Pilling. I walked down the hallway to talk with Captain Ken Heimgartner, Admiral Lopez's EA. I told him I was negotiating for orders and that there was a job in Brussels, but the detailer said he would not consider me for it without backing from a Navy Flag Officer. I finished with, "Ken, would Admiral Lopez support me for this job in Brussels?" He said the Admiral would support me and that he would call the detailer and tell him that. He paused for a moment and then asked, "would you consider being the Commanding Officer of Naval Support Activity (NSA) Naples?" It took a few seconds for my brain to process that request. Did he really ask me if I wanted to be a CO? How does that happen? Can that happen? Close your mouth before a fly gets in. "CO of Naples," I said. Ken said Yes, the CO is about to be relieved, and we need a replacement. Would you consider doing it? I reminded him that I had turned down major command, and he said not to worry about that. I asked if I could get back to him with an answer the following day, and he said yes.

At home that night, Linda and I talked about the opportunity. I had been home from IFOR for a few months, and so our family situation had settled down. The prospect of staying in Naples was daunting. The school was not the greatest. And living in Naples had been difficult for the family while I was away in Bosnia. Every house in our Parco but ours had been broken into. Most of the Americans that lived around us had been robbed. On the other hand, we would not have to move from Naples. We had established ourselves in the community and therefore knew who our friends were and who would obviously be angling to influence the new CO. We had been involved in the NATO activities in Naples but really had not done anything with the folks at NSA or the Navy. I was completely unaware of the pending relief of the CO. I went in and told Ken that I would be delighted to accept the position if Admiral Lopez thought I could help.

Accepting the offer of the job was the easy part. Finally, getting to the job would take some time and effort. The Bureau of Naval Personnel (BUPERs) initial reaction to the mention of my name and command was he turned down major command twice and, therefore, cannot be considered again. Turns out Dodge (Diego Garcia) and Gitmo (Guantanamo Bay) were separate offers. Also, the prospect of rewarding someone who had turned the Navy down did not sit well. Eventually, they got over that, and then the issue became does he have to take over in Naples. Could he go to Rota, Spain, instead? LTG Carter was the guy who was tasked to ask me about Rota. I said, what do you think? He said, "command is command, take it while you can." So, I was now in the running for two commands. The P-3 community allegedly wanted their guy to take over in Naples because he had been on the Commander Fleet Air Mediterranean (COMFAIRMED) staff and that I should go to Rota because no one on the Navy side really knew me in Naples. Eventually, Ken Heimgartner said the CINC is considering you for both jobs, but in the end, you will be assigned to Naples.

Meanwhile, I had been sworn to secrecy, so other than my wife and my boss, I did not think anyone knew about my being in the running for the job. So, we had not even told our daughters about it. One day at Naples American High School, where Linda was substitute teaching, one of the administrators said something about my becoming CO. Linda did her best to change the subject without confirming the information but came home and told me we had to talk with the girls before they found out from their friends. The girls took the prospect of staying in Naples well. I told them my being CO would change their lives and that they would be under a lot of scrutiny. We told them we would be moving to the CO's house and would not have to worry about being broken into and that we would have fans and air conditioning, something that we did not have in our current home. They agreed to stay and not let on to any of their friends know that I would become the new CO.

Finally, it was agreed that I would become the new Commanding Officer of Naval Support Activity Naples, Italy.

The change of command was the shortest and simplest I had ever seen. The Interim Commanding Officer of NSA Naples, Captain Joe Grosel, and I met in the office of Commander Fleet Air Mediterranean (COMFAIRMED) RADM John Ryan. RADM Ryan asked us if we were ready. I turned to Joe and said, "I relieve you, Sir!" Joe replied, "I stand relieved!" I was the new CO. We shook hands, and the change of command was over. Joe could go back to concentrating on his full-time job as Chief of Staff for COMFAIRMED. Although we had not known one another before, we were both from Cleveland, and his support would be invaluable in the future.

COMMANDING OFFICER NAVAL SUPPORT ACTIVITY
NAPLES, ITALY

The sad news is that no one wants to step into command when the former Commanding Officer has been relieved for cause. The good news, in my case at least, was that the relief was not because the command had been doing poorly but was due to the former CO's personal conduct.

My first Executive Officer was Commander John Newcomb, who was succeeded later by Matt Sturges. I had met John briefly when I was the maintenance officer in HSL-37 in Hawaii and the LAMPS MKIII helicopter detachment he was on out of San Diego shore-based at Barbers Point. I wound up getting an hour in the SH-60B in the local pattern. In the small world realm, John's OINC was Larry Larson, who I had known in flight training, and who had been with me at AFSOUTH. Ironically, Larry left AFSOUTH and eventually became Commanding Officer at NAS Guantanamo Bay, Cuba. Having an XO with a similar background helped to ease the transition. John was very conscientious and community-oriented and looked out for our people and the Naples community. I did not know any of the department heads, but the Operations Officer, CDR Mark Vollmer, also was a LAMPS MKIII pilot who was instrumental in establishing the C-12 program in Naples.

The Command Master Chief was initially Master Chief Machinist Mate (MMCM) Randy Ertzberger, who was succeeded by Rick Banning. The Master Chiefs always kept me honest and, like the XO, made sure I understood the ramifications for people that would come from my decisions. I had an open-door policy, and the XO and CMC would wander into my office throughout the day. They both knew I wanted to know what their opinion was on any given issue and not what they thought I wanted to hear. They were all particularly good at letting me know when I was getting off course. In general, our days ended with us getting together in my office over a cup of coffee, going over the day's events, and formulating a plan of action for tomorrow.

As mentioned before, the command was operating well, and that gave me time to get up to speed on the significant issues that I faced. The challenges during my time as CO included: the Naples Improvement Initiative (NII), Embassy bombings in Africa, support for combat operations in Kosovo, care and feeding of flag officers, disaster relief in Quindici, Italy, and involvement in the Department of Defense Dependents School. The good news was that I was the boss, and I would get to fly again!

My responsibilities as Commanding Officer, Naval Support Activity, Naples, Italy, involved providing administrative services, material, and quality of life support to NATO's Allied Forces Southern Europe, Commander in Chief U. S. Naval Forces Europe, Commander Sixth Fleet, Commander Fleet Air Mediterranean, U.S. Army, and U.S. Air Force personnel, and 104 tenant commands. Working with me were 44 Officers, 373 Enlisted Personnel, 270 Department of Defense Civilian Contract Workers, and 554 local nationals with a station budget of forty-five million dollars. These folks were spread out at three sites in the Naples area: Agnano, Capodichino, and Gricignano Aversa.

The biggest challenge was managing the Naples Improvement Initiative. This was a 600-million-dollar project that called for recapitalizing and reducing the U.S. Navy footprint in the Naples area. The overall concept was to transition out of the Agnano area, where the facilities were the oldest and had been damaged by earthquakes. We would relocate the operational facilities to Capodichino, where the Naples Airport is located, and build a new base in the Gricignano Aversa area to the northeast of Naples. Construction at Capodichino would consist of administrative buildings, command, control, communications, computers, and information systems or the C4I facility, Mess Hall, Bachelor Enlisted and Officer Quarters, Navy Exchange, Gymnasium, and a medical clinic. A Naval Hospital, high school, grade school, Navy Exchange, Navy Lodge, and over five hundred housing units would be constructed at Gricignano Aversa.

The job was to maintain the facilities in Agnano while transitioning to the other two sites and trying to cause minimum disruption in a process that was guaranteed to cause massive disruption. Managing people's expectations and trying to alleviate their frustrations during this highly disruptive process would consume all my time. RADM Ryan made it clear to me that my job was to make this happen with the least amount of pain, and he expected me to be on top of the entire process. He cited a project in Rota, Spain, where the BEQ had been built but was sitting empty because they did not have any beds. He wanted me to track everything.

There was a Naples Improvement Initiative (NII) working group that met regularly and that RADM Ryan chaired. The various parties would outline their activities, report progress or difficulties, and await guidance on what their priorities should be. Initially, the Admiral attended all these meetings. Over time, he became confident in my ability to manage the process. I would chair the meetings and then supply updates to the Admiral in his office, or I would pass the information along to his Chief of Staff, CAPT Joe Grosel. These meetings were key in our attempts to minimize the disruption that construction inevitably causes. My job was to be the people's advocate. The engineering solution is always to do what is easiest for construction. That saves time and money. It also usually causes tremendous inconvenience for folks who are trying to work in the facilities that are located around the construction site. I tried to strike a balance between what the engineers and construction folks needed, with what would make the entire process tolerable for all the folks in the community who needed to get to work with a minimum of hassle and inconvenience.

To the extent that I could influence construction on the NII, I tried to go big and go modern. We began to talk about additional administrative structures at Capo, and I wanted them to be as large as possible within the potential existing footprint. Construction would be limited by available area and proximity to the airfield's runway. This was our one chance to get it right, and I did not want folks to say in three or four years that we should have gone bigger. I also wanted to make sure that whatever we did took into account the idea that, at some point, the facilities could be modernized and expanded. I pushed for fiber optic cabling in the buildings even though it was more expensive and not the standard at the time. My mantra was ten years from now, I want people to look at Capo and the Support Site in Aversa and say this was done right. I also suspected that we would have more tenant commands that would move to the area after the NII was completed. At the time, there were discussions about moving the U.S. Naval Forces Europe Headquarters from London to Naples. This would eventually take place in September of 2005 and move into those spaces that I was doing the conceptual planning for in 1999.

My first big decision was whether or not to sign a one-year lease contract for the existing grade school and high school facilities in Agnano. New school facilities were being constructed in Gricignano Aversa, and there was some doubt about whether they would be completed in time for the coming school year. The contract extension on the existing schools had to be signed in the spring. Completion of the new schools was scheduled for the fall. If I signed the contract, then there was a great likelihood that we would have a grade school and a high school sitting empty somewhere. If I did not sign the contract, and the new schools were not finished, then we would not have a place to hold school in the fall. I met with the school folks, the construction folks, and the concerned parents. Fortunately for me, Captain Dan "Gabby" Gabriel, the Chief of Staff to STRIKEFORSOUTH, had been involved in the project from the start as a concerned parent. We talked it over, and his recommendation was to go all in on the completion of the new schools. He smiled and said, remember when

Alexander burned his ships, and the troops did not have any options but to fight? Don't sign the contract, and the only option is to complete the schools. It was a near-run thing, and I was first introduced to the term soft opening, but the schools were open, but not entirely finished, for the fall.

Military Personnel used to refer to me as the mayor of Naples. I had resources, responsibilities, requirements, and expectations. Balancing between all of them was a constant juggling act. One of the fun aspects of the job was doing the Area Orientation. All Personnel who were stationed in the Naples area were required to attend. Instead of having them wander all over the bases to get their check-in process completed, we brought the process to them in one place. RADM Ryan or CAPT Grosel would address the group and welcome them to Naples. Then I would come on stage and outline what we would be doing for the day. I had several stock lines that I used in the presentation that you have seen before.

<u>Nothing in Naples is easy, but everything is possible</u>. There was always a way to get what you wanted to get done. You just might have to go in a roundabout way to achieve it. The initial reaction to most requests was, no, that cannot be done, it is impossible. Over an expresso, with a little cajoling and a sense of, if you do this for me, I will do this for you, and things will happen. Underlying all of this was the idea that there would be payback at some point in time. If I help you now, then when I come to you in the future, you will be obligated to help me. And there will be a time when I will come to you for the return favor.

<u>Choices have consequences</u>. I would tell the assembly, I cannot solve all your problems or make your choices, nor do you want me to do so. What I can do is frame your options or available alternatives and make a recommendation to you. But I cannot make

the choice for you. So, if your choice does not pan out, you must deal with the consequences, not me. However, we would do everything we could to lessen the consequences.

Naples is like a beautiful woman with dirty feet. In the span of a hundred feet, you could see something like a gorgeous statue or scenic vista and then look down at a pile of garbage on the street or a burning dumpster. You had to make the mental adjustment and focus on the good while dealing with the bad. Over the time I was in Naples, the city made a concerted effort to clean a lot of the grime and dirt up with good success, but you would still be exposed to some of these jarring contradictions while going down the street. There were also the "campfire girls." Not the kind back in the states. These were working women who would stand around a fire in a fifty-gallon drum and wait for their customers to pull up. At Christmas, they would dress up in the proper colors for the season. The kids of one of the guys I worked with asked who those women were, and he replied, they are Santa's elves.

The Neapolitans will give you the shirt off their back and then steal it. As mentioned previously, there was a tremendous amount of crime. In Lago Patria, where my family initially lived, everyone in our Parco was robbed. All the Americans in the vicinity also had their homes broken into. We did not. The only thing we could think of was that we had dogs, mixed breed, but primarily German Shepherds.

At one point, when we were living in Lago Patria, our neighbor's duplex was broken into. We were house-sitting for them, and the day before they were supposed to return, I went over to check on their place. When I went up to the door and started to put the key in the lock, the door opened. It was dark, so I went over to our house, got a baseball bat and the dogs, and entered the home. The dogs ran through the house, and I knew there was no one inside.

When I turned on the lights, it was apparent that there had been a burglary, but the burglars had been disturbed. They left several valuables, and the window that led to a vineyard behind our houses was open. They had been startled by something and then rushed out the back. I called Navy security, and they came out to take a report. They agreed that the thieves had been disturbed and asked if we had seen anything. I had to say no. A couple of weeks later, we were discussing the whole incident over dinner and were still trying to figure out what might have happened. Our youngest daughter said that she and a friend who was staying overnight had engaged in a screaming contest while we were out at a social engagement. Who knew that the best security system was a couple of teenage girls?

The bombings of the American Embassies in Kenya and Tanzania in 1998 changed my whole outlook on security. Up until then, I had been working under the assumption that terrorist organizations would refrain from striking U.S. targets that were embedded with the local nationals because of the possibility of killing or injuring the local populace and thereby turning them against the terrorists. Africa changed all that for me. Our force protection posture was increased, and we stayed at an elevated level for the duration of my time in command. We reviewed all our installations and facilities to figure out their vulnerabilities and how we could minimize the danger to them. We had a security force made up of our Master at Arms (Sailors), the Marine Barracks (Marines), and an Auxiliary Security Force (ASF - also Sailors assigned to tenant commands, not NSA Naples). The MAA and the Marines did security for a living. The ASF were sailors from the tenant commands who were identified and trained to augment our regular security forces in times of crisis. This would usually be for a specific event like the Fourth of July or when we ran security drills. In this case, we activated the entire ASF. That meant those individuals were now working for me. When their commands asked, how long will this last? I said indefinitely.

After doing our security assessment, we decided that some activities should be closed while others would now have an increased security presence. All our security folks were in full battle rattle and armed. The idea of withdrawing from "soft" facilities, which could not be protected, into "hard" positions, where we could provide force protection, became known as Operation TURTLE. We were going to withdraw into our shells and make it hard for someone to hurt us. All the decisions were subjective, and not all of them were popular. But at the end of the day, they were the ones that we could defend. As my security officer said, "Skipper, you cannot stop all terrorist attacks. The best you can do is make it as hard as possible for them to get at you and hope they will go for an easier target somewhere else. At the end of the day, if they are determined, they can get at anybody, anywhere."

We ramped up the security, and I used to roll the dice to decide the order of inspection for vehicles entering the base so that it would be entirely random. It did not matter who you were. If your number came up, you were inspected. I cannot count the number of times I would be driving onto the base, and the Marine or Master at Arms would smile and say, "Skipper, your number came up, pull into the bay so we can inspect your vehicle." One day I asked the MAA, do you know why I pull over for inspection when my number comes up?" He said sure, that way, when someone complains about us conducting an inspection on their vehicle, we can say the CO of the base does not complain when we do it to him. What is your beef?

One day I was on my way into Agnano, and the traffic was stopped going into the base and was rapidly backing up. I hit my blue lights, pulled up onto the sidewalk, not unusual in Naples, and got up to the gate. When I asked what was going on, the MAA said there was something suspicious in the men's room at the Navy Exchange. The OOD came up and said that they had been using the bomb dog to sweep the Navy Exchange and Commissary and that it had alerted them to something in the men's room. He said the Italian bomb squad had been notified, but they could not provide an estimated time of arrival.

I went into the men's room, and there was the dog, the dog handler, and some of the security folks. The handler reported that the dog had alerted a couple of times and had focused on the paper towel dispenser located on the wall. The MAA said they could not see anything wrong but called it in because of the dog alerting to the dispenser. I asked for an update on the arrival of the bomb squad. They said it would be a couple of hours at best. I looked at the MAA, the handler, and the dog and asked, "do you feel lucky?" I said I did not think there was a bomb, and I was unwilling to tie everything up for hours pending the arrival, or not, of the bomb squad. So, I was going to knock the towel dispenser off the wall, and if anyone had any objections, they should get out of the men's room right now. They said, "Skipper if you are staying, we are staying." A deep breath, a striking blow, and the dispenser flew across the room, and there was no explosion. Later that day, when the XO called and said he heard there had been a bomb threat, I laughed and said, "John, you don't know how close you came to major command!"

The situation had been deteriorating in Kosovo over time, and it became apparent the U.S. would respond militarily. We began to see increased activity on our ramp as aircraft would divert into Naples to get fuel. Capodichino Airport was a dual-use facility. The Italian Air Force was responsible for the field and air operations, there was also a commercial area at the airport where regular civilian flights took place, and then there was a U.S. Navy area where we provided support to transiting U.S. aircraft. I went to talk to the Italian Colonel in charge of Capodichino about providing support for the forces engaged in the Kosovo campaign. He reminded me that the U.S. Navy were guests on the facility and that he required notification in advance of potential operations so that he could keep his military and civilian superiors informed about what was going on. I told him I would do my best to ensure that when I received any information on operations, he or his people would receive an immediate phone call to let them know what was happening. Again, it was the old adage, "no surprises."

At times, our ramp space was full of transiting aircraft, and there was a constant stream of fuel trucks. Depending on weather and operational schedules, we were constantly surging, sorting, and scheduling support operations to some exceptionally large and thirsty U.S. Air Force aircraft.

The fact that the U.S. was conducting air strikes in Serbia also did not help our security situation. There were significant numbers of folks in the Naples area that came from the Balkans, and we had to increase our vigilance and force protection posture to convince anyone who might be watching that we were not a soft target.

The most interesting Kosovo event occurred when my secretary said I have an Army Colonel out here who wants to talk with you immediately. The Colonel introduced himself and said that he was in charge of a group of Army Apache AH-64s that were trying to make their way over to the Balkans but had run into dangerous weather and were looking for a place to spend the night. We figured out how many aircraft were coming, and I told the Colonel we would do everything we could to support him. My ramp space is limited. I never know how many or what size USAF aircraft are coming. I will have to pack your aircraft into a pretty small space. We have folks on my staff who do this for a living on ships, and if your folks would cooperate, we could make this work. I asked, where are your people going to stay? He said he had not had a chance to make arrangements for their billeting. I said I can give you rooms in the BOQ and BEQ, and we can put the overflow on cots in the gym. We will open the CIAO (Mess) hall for 24 hours so that your folks can have a place to rest and get something to eat. Once the weather clears and you are gassed up and ready to go, then we'll get you launched and on your way. He thanked me and said this was the first time anyone had gone out of their way to help him and his folks during this whole evolution. I pointed to the wings on my chest and said we are all in this together.

My most delicate job was the support of the numerous flag officers who were stationed in the Naples area. I was their landlord. CINCUSNAVEUR, COMSIXTHFLT, and COMFAIRMED all were in the area and had daily access to the bases. So, I could see my boss, my bosses boss, and my bosses, bosses, boss anytime, anywhere. I was responsible for the maintenance and upkeep of their quarters. In addition, I was the landlord for an Air Force Three Star, an Air Force Two Star, a Marine One Star, an Army Three Star, and an Army One Star. My Public Works Officer, Captain Jim McConnell, and my Flag Liaison Officer did a fantastic job of running the program. I knew most of the Flag and General Officers from my time as the EA to the Chief of Staff at AFSOUTH, so I had a fairly good idea of the ones that would require more attention. My quarters were in a duplex located on a gated compound in Pozzuoli. They had originally been occupied by an Air Force Three-Star, so there was a generator, safe room, enhanced communications, and a few more upgrades, as you would expect in a Three-Star's quarters. In addition, the other side of the duplex was occupied by the EA to CINCUSNAVEUR, Captain Heimgartner for Admiral Lopez, and later Captain Don Loren for Admiral Ellis. A lot of potential issues were diffused by casual conversation with the CINC's EA at our quarters. The good news was that COMFAIRMED told me on the first day on the job, "it is your base. You run it the way you think it should be run." There were occasional suggestions from RADM Ryan about how we could make things better, but he handed me the keys to the command and let me do the driving.

I think the first time I heard about a mudslide in the Italian town of Quindici was when the American Consul General in Naples called my office. Quindici was a small-town east of Naples with about 3,000 residents. Mudslides had started in the area following heaving rains. The first report was that up to two hundred people were missing and feared dead. The Italian government was mobilizing a rescue effort. The Consul General said the Italian Government had asked her what aid the U.S. could provide. I told her the request would have to go from the Department of State to the Department of Defense and then come down from EUCOM to CINCUSNAVEUR. In the interim, I would begin preparations to help the Italians in any way we could. The first request had to do with body bags and setting up a camp for rescue workers. I called the hospital and asked about the body bags, and I knew that would not be the last thing we were asked to do.

I set up a meeting with Captain Jim McConnell (Public Works Officer) and Major Jim Chartier, the CO of the Marine Barracks. The Public Works Officer was a Navy SEABEE. He had a lot of experience in building and destroying things and said his SEABEE's could supply a lot of help, but we had to do an assessment of the area and the damages first. The Marines could supply search and rescue aid and had the communications and experience of working and living in the field. I asked them to get ready and said we would try to get into Quindici as soon as possible. I believe the next thing that happened was Admiral Lopez calling to ask me what we could do to help. I gave him a brief run down and said I was waiting to get U.S. and Italian clearance to go to the city. He told me to get on the road at once. The Consul General, Captain Jim McConnell, and I went to Quindici. We met with the local officials, offered our condolences, and asked how we could help. We told them we were prepared to construct a temporary camp. Captain McConnell outlined the equipment and capabilities that we would be glad to provide in their hour of need. We also offered the

Marines to aid with the search and rescue effort or any other tasks which they had for us. They welcomed our offer, and I contacted the local Italian Military Officials who were coordinating the effort.

I had not expected the media interest in the tragedy of Quindici. I also took my Public Affairs Officer (PAO), Lieutenant Jeff Gordon, along with the Consul General and Captain McConnell. He was fielding inquiries from the press about the U.S. aid that was being provided. He spoke fluent Italian, so he was able to conduct several interviews with the Italian press so that we could get out ahead of the story. At one point, he told me that one of the networks in the U.S. was on the phone and wanted to do an interview. Having a radio/television journalism background, I was prepared for the questions and knew what I wanted to emphasize. One, our thoughts and prayers were with the residents in Quindici and neighboring towns. Two, we were responding to requests from the Italian government through our State Department for assistance. That assistance would be in building a camp using our SEABEES and Marines and whatever other services we could provide to the Italian authorities coordinating the relief effort. Three, the level of devastation in the town was tremendous. I had never seen a mudslide before, but I had seen what lava could do in Hawaii. I said the force and volume of the mudslides were tremendous. The mud would enter through one side of a building, through a door or a window, and would blow out the back walls.

After we had toured the city, we returned to the base in Naples and roughed out the details of how we would help. The SEABEES would set up the rescue workers' camp and some tents for the SEABEES and Marines who would provide assistance. I also said we should put out the word for volunteers from the base, and we would need to arrange for transportation and billeting for them. I told the XO I wanted a situation report to go out every four hours for the duration of our involvement. I said I wanted to push information up the chain of command, so we were not being micromanaged by higher authority.

I made it a point to ensure that the CINC's EA and the COS for COMFAIRMED were aware of what we were doing. At one point, the XO asked me, where did you learn to manage a crisis like this? I thought for a moment and said it was the time I had spent as the EA to the COS LTG Carter at AFSOUTH. I learned to be comfortable communicating with Flag Officers and that, in general, the more information you gave them, the less inclined they were to get in your business. As a Junior Officer, I had always believed that when you were talking with your senior, one of two things could happen, and one of them was not good. So, I was reluctant as a JO to interact with senior officers. As I gained experience, I realized that my seniors appreciated someone who would keep them informed and would not hesitate to tell them the good or the bad. You got a lot more help both in the form of suggestions and resources when you were forthcoming with your superiors in the chain of command. That does not apply in all cases, and my situation may have been unique in that I had a personal working relationship set up with CINCSOUTH and COMFAIRMED. There was never any second guessing or requests for additional information from folks in the Chain of Command, including EUCOM in Stuttgart.

At the end of that first day dealing with Quindici, I was exhausted, and when I got home, the first thing I did was make a rum and coke. Then the phone rang, and it was the PAO saying that CNN really wanted to do an interview. Skipper, I know that you have been doing interviews all day, but could you do one more? I said fine. He said the network would call me in about ten minutes before we were to go on air and that we would be the lead story on the broadcast. About fifteen minutes before I was supposed to go on air, Linda called. She was on a school athletic trip with the

Naples American High School students in Italy and was checking in. I said, "I cannot talk to you right now, CNN is going to call, I'll call you later," and then hung up the phone. The producer called shortly after I hung up with my wife, outlined the process and thanked me for coming on. At the top of the hour, I went live on CNN. I was sitting in my bedroom, sipping a rum and coke, looking at the screen that had my name on it, and going over our disaster relief efforts. Surreal does not begin to describe what I was feeling at that moment.

As the days went by, we settled into a routine in Quindici, and I began to wonder when our help would no longer be needed. I visited our folks occasionally and would talk with the Public Works folks daily. I knew that getting into crisis management was easy. Getting out, well, that was a whole other thing. As the level of our aid declined, I surfaced the idea with my chain of command, the Consul General, and the Italian Military Officials on setting up an end date for our assistance. We were able to begin moving people and equipment back to Naples, and our aid ended.

After it was all over, I thought about the fact that I had never received any inquiries from the chain of command or second-guessing on our operational and logistical aid. I also had not received any guidance on whether or what I should say to the media. One day the PAO walked into my office and asked if I had given an interview with one of the leading Italian newspapers. I said no, I always go through you. He put a paper on my desk from one of the leading Italian newspapers. My Latin was a little bit rusty, but I looked up and said, "does it really say that I recommended to the Italians that we use Napalm to break up some of the log jams in the streams around Quindici? " He laughed and said, I figured you didn't say that. You never know how close to disaster you are.

The most surprising thing about my time in Command in Naples was my involvement with the Department of Defense Dependent Schools (DODDS). I found myself involved in a host of activities that were not mentioned when I attended Major Command Ashore School in Washington, D.C. It started with the decision on whether to sign the lease for the grade school and the high school. I also became involved in their disciplinary proceedings. The schools were located on the base, and I had authority to grant or restrict access to the base. I could not expel a student from school, but I could prevent them from getting on base to go to school.

You might ask, why would I want to do that? Because the school system would not discipline their students. Our experience as a customer of DODDS was mixed. Most of the teachers were fine, and some were outstanding. But there were a fair amount who were just going through the motions. That is ok when you know they are coasting and your child can take the course from someone else. In DODDS, you might have only one teacher for a given course, and if they are bad, then your child suffers. If you bring up the teachers' shortcomings, there is retribution. My wife was a substitute in the DODDS system and had taught in Bermuda, Hawaii, and Naples, and would eventually teach in Brussels. She knew who the good teachers were and tried to have our girls avoid taking courses from the bad ones. The administrators' and teachers' attitudes towards her changed when I went from being a nobody in NATO to the Base Commander. They knew I had an alternative source of information, one that predated my time in command. So I was able to ask more detailed questions about school policy.

The schools were out at what we called the support site in Gricagnano Aversa, while the post office, commissary, exchange, and disbursing offices were either in Agnano or Capodichino. The teachers, and the residents at the support site, had legitimate complaints about not having these services readily available. We installed a temporary post office and a mini-mart. The teacher's main complaint was that disbursing was not open when their hours allowed them to get to Agnano. Disbursing did not work for me, but I met with the OINC and worked out a plan to provide extended hours so that the teachers could do their banking. I also asked him to track the usage by the teachers. I had been told that it was not possible for the teachers to get to Agnano during the day because of their schedules. The most vociferous teacher was the union representative, who said we were being unreasonable by not extending hours. About a month after we extended the hours at disbursing, I ran into the union representative coming out of the disbursing office in the middle of the work day. It was not impossible for him to get to Agnano that day. After ninety days, the disbursing OINC came to me and said, Skipper, here is the log for the folks that have used our extended hours. I do not think the answer was zero, but you could count them on the one hand. He said, I cannot support continuing this based on usage. I agreed and told him to shut it down. When I left, the teachers gave me a report card. I got a "D" in support, and disbursing was one of the reasons they cited for my poor grade.

I had always prided myself on my communication skills. Starting out as a broadcast journalist had prepared me to be able to speak and write fairly well. I told my PAO; I am either your worst nightmare or the easiest guy for which you will ever work. As the CO, I would now have my own weekly newspaper column and radio show. Panorama was the weekly newspaper that was available to all the U.S. personnel in the Naples area. I used my column to highlight issues in the community and provide recommendations and solutions for how to manage them. The radio show provided me an opportunity to speak directly to the community on given issues in much more detail. This was especially useful when we tightened the security at all our facilities in conjunction with OPERATION TURTLE.

I also held quarterly "Straight Skinny" meetings with the community to give them an opportunity to talk to me directly about the issues that were of concern to them. All my department heads would be there, and they would each report to the community on the status of issues they were working on. The audience had a chance to question them directly on anything they wanted to. I would then get up and take questions from the audience. You never knew what would be asked or how you would react to the question. I liked having to react quickly and problem-solve in real time with the folks in the audience. The issue was when someone tried to take over the meeting or would not move on to another topic. Several times I had to redirect the meeting when some more senior officers would attend and try to monopolize the discussion or try to intimidate me into going along with their recommendations.

Every day when I arrived at work, I would go to my office and look out the window to the southeast to see if Mount Vesuvius was active. If there was no smoke, it was going to be a good day. The last time Vesuvius erupted was in 1944. I worried about what we would do if it erupted on my watch. Getting around Naples on a "good" day was difficult. Throw in every Italian trying to get out simultaneously, and you would have massive gridlock. The instruction on Mass Evacuation was old and out of date. I said let's run a simulated evacuation. We'll take folks from the support site, process them, and then fly them out on an H-53. It would be on a small scale, but I felt it would be enough to confirm existing procedures or identify those that we had to rewrite. On the given day, we started the drill. Embedded in the evacuees would be my Linda and my daughter Erin. I knew they would provide me with a separate set of eyes and perspectives on how the whole thing went. There were problems and discrepancies along the way, and the H-53 had to cancel because of higher priority tasking, but we learned a lot and began rewriting our plans. We told folks if they could not get to Agnano, Capodichino, or Aversa, they should be prepared to shelter in place, and we would come to get them as soon as we could. We updated the lists of items they should stock, documents they needed to have on hand, and the need to have a current listing of their address and phone numbers. Vesuvius finally erupted again in 2020.

The security folks pointed out to me that there had never been an emergency evacuation of the C4I facility on Capo. It was kind of a dare, as in how serious is the old man about this. The building housed a number of senior staff, including my boss, COMFAIRMED, and the common belief was that I would not have the guts to order the drill or that the Admiral would tell me to think again. Taking RADM Ryan's missive that it is your base, do what you think is right, I talked with the Admiral. He agreed that we should run the drill. We did everything we could to minimize the disruption by making sure everyone knew the drill was coming. As expected, there was pushback from some of the occupants of the building on the day of the drill, but we were able to complete it. Another list of lessons learned and proposed changes to our operational plan came about because we were able to run the drill.

I have always had a soft spot in my heart for Marines. My brother served in the Marines, 1st Battalion, 1st Marines during Vietnam. I thought I had joined the Marines during my first four months in Aviation Officer Candidate School. I now had Marines working for me as the CO of NSA Naples. The Marine Barracks provided security for the base and some of the more sensitive buildings on the base. They also occasionally provided me with some unusual situations to deal with.

Major James Chartier was the Commanding Officer of the Marine Barracks, and we had a good relationship. He never came to me with problems. He came to me with solutions to the problems he was about to bring to me. I had arrived at the quarterdeck, and the OOD told me there had been a minor fire in the Marine Barracks the night before, but there were no injuries or significant damage. As I was getting a cup of coffee, there was a knock on the door to my office. I knew who it was. Jim said, "Skipper, I wanted to brief you about the incident in the barracks last night." The Marine had been smoking in bed, fallen asleep, and the cigarette caused a fire that damaged the mattress and the room. The Marine was not injured. He was not supposed to be smoking in bed and could be prosecuted under the Uniform Code of Military Justice.

Jim asked if I would let him handle the incident. He said the Marine would pay for the mattress, repaint the room, and had "volunteered" for extra military instruction on how to take proper care of his quarters. A couple of days later, when I was walking to the office, I noticed a "pup tent" just outside the Marine Barracks. When I got to the office, I asked the Master Chief, what's going on? He said the Marine involved in the fire had paid for his mattress, repainted the room, and would be living in the pup tent for the next week to remind him of the necessity for taking care of his assigned quarters. The Marine would practice living in the field, in this case, the pup tent in front of the barracks. I laughed and said, I wish we could get away with something like that!

If you have ever dealt with the Marines, you know they take their security seriously. I was reminded of this during my time in command. A couple of NATO officers were attempting to enter the C4I building and were asked to supply the proper identification. There is a marked difference in the attitude of some foreign militaries about the relationship between officers and enlisted Personnel. Put a separate way, in some NATO countries, enlisted men don't tell officers what to do, ever! In this case, the officers ignored the sentries' commands and tried to force their way into the building. The sentry responded by challenging the officers and ordering them to follow the requirement to supply proper identification and halt. When they did not comply, the sentries forcibly put the officers on the ground until they supplied their identification.

The XO of the Marine Barracks was at my office early the next morning as I was getting my coffee and briefed me about the incident. We ran through the procedures and actions the Marines had taken. While the Marines were technically correct in what they did, the XO and I agreed that the situation could have been managed less drastically, and he agreed to review the sentries' instructions with his Marines. The NATO officers' Chief of Staff came to my office complaining about the treatment the officers had received. I said the Marines handle security in the building. In that sense, the Marines were acting like cops. As the son of a cop, brother of a cop, with lots of cousins who are cops, can you guess where my loyalties were? The upshot of the conversation was that I told the COS that if your officers follow my Marine's instructions, then there won't be any problems. If they don't follow instructions, they won't be allowed on the base. Which will make it exceedingly difficult for them to get to work. I said if the officers have a problem with that,, they can come and talk with me. One of my most prized possessions is a K-Bar knife that was presented to me by the Marine Barracks with the inscription on the blade, "From Your Marines."

While Marines are dear to my heart, Navy Chief Petty Officers are the best. I learned at an early stage in my career to talk to the Chief when I had questions or issues. I was the CO of a base because CPOs throughout my career had kept me on the straight and narrow with sound advice and recommendations. I was a firm believer in a CPO Disciplinary Board to review cases and make recommendations about what kind of discipline should be taken with our junior sailors. I valued their advice. In Naples, administrative space was at a premium, but we made sure to set aside space for a "Goat Locker" or CPO Mess where the Chiefs could relax and socialize. The Goat Locker is a Navy tradition and is for exclusive use by the Chiefs. You are not allowed to enter the Goat Locker unless you are invited in by a Chief Petty Officer. Everyone, Enlisted, Officers, and even the Commanding Officer must request permission to enter or be invited into the Goat Locker. When the CPOs had a going away ceremony for me, I had to stop several times and compose myself. I was thinking about my Dad, who had passed away prior to my taking command in Naples, his time as a CPO, and the guidance he had given me to always listen to your Chief. I was especially touched when the Naples Chief's Mess, represented by Command Master Chief Rick Banning, made me an honorary Chief Petty Officer.

The thing I liked least about being in command was holding Captain's Mast.

I did not know it at the time, but the way I would conduct Captain's Mast when I was a Commanding Officer was formed by my experiences with the Uniform Code of Military Justice (UCMJ) while stationed in Bermuda. I hated it when you took a sailor to Mast after you had done everything you could to straighten them out, and the CO would give them a heart-to-heart talk and then dismiss the charges. My policy as CO was that Mast was not a counseling session. Captain's Mast was an admission of failure on the part of the chain of command to successfully modify a sailor's behavior. That meant that the accused sailor's Leading Petty Officer, Chief Petty Officer, Division Officer, Department Head, Chief Petty Officer, Disciplinary Review Board, and Executive Officer had failed to solve this sailor's disciplinary problem.

I relied heavily on the CPO Advisory Board to function as a filter to identify those sailors who had the potential for further service vice those who should be put on a path to being discharged from the Navy. I felt that XO screening, which preceded Captain's

Mast, was the place for sailors to receive an "ass chewing" if that is what the CPO Disciplinary Board felt was appropriate, as opposed to the sailor being taken to Captain's Mast. If everything else failed to get a sailor's attention and straighten them out, then it was time for Captain's Mast. For me, Captain's Mast was about assigning punishment, not additional counseling. If there was any doubt in my mind about the sailors' potential for future service in the Navy, I would max them out on punishment (fine, reduction in rank and confinement, or extra duty) and then suspend the fine, reduction in rank and confinement, for six to nine months if they could keep their nose clean. My parting words were always, "I don't want to see you in here again. If you come back to Mast, the suspension is vacated, which means you will enter one rank lower. Then I will reduce you another rank, and you will receive the maximum punishment that I am authorized to give."

The stupidest thing I ever did, as far as the Military Justice system goes, was when I threatened to hold Captain's Mast in front of the entire squadron in HSL-37. We were at quarters, and there had been a couple of Driving Under the Influence (DUIs) lately, and we were not getting through to our folks about how serious this was. We encouraged our sailor's to use the buddy system or designated driver. We also had the "Tipsy Taxi," where a sailor could call a designated Taxi service and would be given a free ride home. There was no good reason for us to keep having the DUIs. At quarters, I said that the next sailor that got a DUI would be taken to Mast before the entire squadron. I was a little wound up, and my mouth got ahead of my brain. Up in my office afterward, the XO, Dave Taylor, asked, did you really mean that? I said that it sounded like a good idea at the time, but now, not so much. He said it would not look good if some airman started acting like a "sea lawyer" and giving me a tough time in front of the entire squadron. I could not take the statement back, but I dreaded that someone would get a DUI, and then I would have to make good on my promise.

Fortunately, we did not have a DUI for the remainder of my time in the squadron.

The best thing about being in command is the people you get to work with. The next best thing is when the job involves flying. Being in command in Naples meant that I would be able to go flying again. Naples was setting up a C-12 program. The C-12 is the military equivalent of the Beech King Air civilian aircraft. We would have two planes and contractor support for their maintenance. All Unrestricted Naval Aviators (Helicopter Pilots) learn to fly fixed-wing aircraft before they are taught how to fly helicopters. The last time I had flown a fixed-wing airplane was in 1974. Now I would be flying a multi-engine fixed-wing aircraft. I went to Wichita, Kansas, to learn about the aircraft and train in the C-12 simulator. My biggest challenge was learning to fly with my left hand. The right hand had about 3,500 hours of flying an aircraft with a stick. The left hand had 0 hours with a yoke. What that means is helicopter and jet pilots fly using their right hand on the stick. ASW and transport pilots fly the aircraft with either hand using a yoke, depending on which seat they are in. The other hand is on the throttles. I managed to learn all that and returned to Naples and completed a local flight training syllabus.

My intention initially was to become an aircraft commander and fly as much as I could. I used to look down on those senior naval aviators who were not current in model and did not hold all the available qualifications. I soon learned that there were not enough hours in the day to be Base Commander and Ace of the Base. The Aircraft Commanders, who were all junior to me, were great about indulging me on the days I was their copilot. Normally, the copilot would do all the work of planning the flight, filing the flight plan, and doing the pre-flight inspection. Then the Aircraft Commander would conduct the briefing. I tried to be a good copilot and perform my required duties. Things, like running a major overseas Naval

Installation, tended to get in the way. After a while, the Operations Boss Commander Vollmer said, "Skipper, just make the briefing time. We will take care of the rest." When I could, I did do all the copilots' duties, but that was about 15% of the time. I loved the flying and would usually take the weekend hops because that would not significantly impact my time in the office. But I was not able to gain the systems knowledge and flight experience that would allow me to actually become an aircraft commander.

I flew the C-12 like a helicopter pilot. Constantly waiting for something bad to happen or for something to go wrong. Also, I was a little uncomfortable at flight level 300 (30,000 feet) and above. For a helicopter pilot, thirty thousand feet is a long way to fall. Then there was the autopilot. I had never flown an aircraft with an autopilot that could get me from point A to point B, where I did not have to touch a thing. The first time the aircraft commander pulled out a newspaper and began to read when we reached cruising altitude blew my mind. Most of the C-12 pilots assigned to Naples were Lieutenants. They had come from multi-engine communities flying the P-3 and E-6. They laughed at their CO, who wanted to "hand fly" the aircraft as much as possible instead of letting the autopilot do the work. Like most naval aviators, I wondered what it would have been like to fly for the airlines. Flying the C-12 all over the Mediterranean gave me a chance to get that experience. I never felt so grateful for having flown the H-2 and spent the majority of my flying career as a LAMPS MKI helicopter pilot.

My last flight as a Naval Aviator was on 1 June 1999. We flew from Naples to Tirana, Albania. The flight was a little over three hours long, and half of it was at night. I remember distinctly thinking as we took off from Tirana to Naples that this would be the last flight of my Navy career. I took off from Tirana and landed in Naples, hand-flying the plane all the way back, including the instrument ILS approach. CDR Vollmer gave me 3.0 hours of first pilot time. He took .01 as the copilot. My logbook was closed out. I had flown 3,681.9 hours during my career as a Naval Aviator. In six flying tours, I accumulated 1,043 shipboard landings with 287 at night, almost a thousand hours in actual or simulated conditions, and almost seven hundred hours at night. The two aviation goals I did not reach were to accumulate 3,000 hours in the H-2 and to go over 4,000 total flight hours. I had a full and challenging aviation career, beginning as a Search and Rescue Pilot in Bermuda, moving on to the tremendously challenging and satisfying assignments in the LAMPS MKI community, and culminating as a multi-engine transport pilot for my final flying tour.

Being the CO of a major overseas installation was different than being a Squadron Commander. At Major Command Ashore School, they used the following analogy. You were the equivalent of an all-star baseball player when you were a Squadron Commander. Now, as a Major Base Commander, you are going to coach the soccer team. In most cases, you were dealing with areas that you had relatively little experience with, such as Public Works, Supply, Security, and Base Administration, etc.. I had been fortunate enough to serve at a Naval Air Station at the beginning of my career and had a feeling for how these departments worked within a larger shore-based organization.

As always, the time seemed to fly by, and before I knew it, I was looking at leaving Naples and considering where our next adventure would be. I did not believe it would be in Europe, but it was.

CHAPTER XVII

AT THE TOP OF NATO

INTERNATIONAL MILITARY STAFF BRUSSELS, BELGIUM

MILITARY ASSISTANT

TO THE DEPUTY CHAIRMAN OF THE NATO MILITARY COMMITTEE

JUNE 1999 – JUNE 2002

My first thought was that I would be returning to the states following command in Naples. We had sent two daughters off to college from Europe and the youngest would be entering her senior year in high school. We believed she wanted to return to the states and doing so would better position her for going to college. At some point, she said she would not mind remaining in Europe for her senior year in high school. We began to think about a NATO job in Belgium.

Despite having a successful tour as a Major Base Commander, I was under no illusions about my future. My chances for Flag rank were zero; no one in the LAMPS community would be pushing for me to be selected and BUPERS would not forget that I had initially turned down major command. I had collaborated with the folks at the Supreme Headquarters Allied Power Europe (SHAPE) extensively when I was stationed at SACLANT in Norfolk, and also while I was posted at AFSOUTH in Naples. I had no desire to live in Mons, Belgium. Duty at the NATO Headquarters in Brussels, Belgium offered more opportunity- I wanted to be at the place where the strategic guidance and direction originated, versus being at the end where they were executed operationally. In effect, I wanted to be where the "strategic sausage" was being made. There were three potential staffs in Brussels on which I could work. There was the U.S. Mission, the U.S. Military Delegation to NATO, or the International Military Staff. The U.S. Mission (The Mission) were the State Department diplomatic folks at the NATO HQ including the U.S. Ambassador to NATO. DOD had a representative within the mission and there were military officers who supported him. Then there was the U.S. Military Delegation to NATO (The Delegation). This was the U.S. Military Representative (MILREP) to the NATO Military Committee and his staff. Last, there was the International Military Staff or the IMS. The IMS supplied support to the NATO Military Committee and there were a number of U.S. military personnel on that staff.

The Detailer was happy to hear that I wanted to go to NATO and gave me the number of a U.S. Army officer who managed the assignments in Brussels. The Colonel said:

"I have two jobs that look like they would be a good fit for you if you are interested. One is on the U.S. Military Delegation as a Navy Planner working for the U.S. MILREP. The other is as the Military Assistant (MA) to the Deputy Chairman of the NATO Military Committee (DCMC) on the International Military Staff (IMS)." He gave me a point of contact who was working for the DCMC and I gave him a call. He described the respective jobs, and it turned out that he had served in both. I asked him, "which one would you recommend?" He said the MA to the DCMC. The folks working for the MILREP were subject to phone calls at all hours from the JCS, worked weekends, and it could be pretty hectic. The MA job was more laid back, observed NATO working hours, and didn't deal with the Joint Staff.

Having come from consecutive tours as EA to the COS at AFSOUTH, and a command tour as CO NSA Naples, I was looking for something a little less hectic. As a result, I asked to be put in for the MA job. The current DCMC had previously been the U.S. MILREP. Lt-Gen Mike Byron was a Marine Three Star who chewed tobacco, spoke fluent Spanish, and could dance a mean salsa. He was also an expert diplomat when it came to working the members of the NATO Military Committee. However, I did not know any of that when I asked to be nominated for the position. I got a call a few weeks after my name was submitted and was informed that I had been accepted for the position.

After a few months on the job, we were in the office and the DCMC brought up my selection as his MA. "Have you ever wondered why you got the job?" he asked me. His request for an MA was based on the following criteria: Senior O-6, post major command, previous NATO experience, no young children, and operational background in combat arms. As he was talking, I thought to myself, ***check, check, check, check, and check***. He said the other nominee was an Army Lieutenant Colonel who had been selected for Colonel, had never been in command, first NATO tour, two young children, and was an intelligence officer. General Byron smiled and said, "tough choice, huh"?

The DCMC's staff included the MA, Aide, Driver, Enlisted Aide, Yeoman, and a civilian Administrative Assistant. Our schedule revolved around that of the North Atlantic Council (NAC). The NAC met in regular session at the ambassador level on Tuesday for an informal luncheon, and formally on Wednesday to make decisions. The Military Committee met on Thursdays. The DCMC met with the Chairman of the Military Committee (CMC) on a regular basis during the week. These meetings would take place every morning or afternoon as arranged by the Director of the International Military Staff (DIMS), a three-star officer. When I was the MA to the DCMC, the DIMS respectively were a Lieutenant General Danish Army officer and an UK Royal Navy Vice Admiral.

The DIMS was in charge of the IMS, in effect the Chief of Staff. The DIMS was in daily contact with the folks on the staff at SHAPE in Mons, Belgium; and supplied them information and background on what was going on in the NAC and the MC. This provided SHAPE with an idea of the sort of questions that would be coming from Brussels on a variety of issues and the opportunity to begin laying out proposed responses or a plan of actions and milestones.

The DIMS also set the battle rhythm for the CMC and the DCMC. His staff provided the papers and recommendations to the CMC based upon the input received from the staffs of SACLANT and SACEUR. Monday's meeting was typically held to review what had happened over the weekend, review the schedule for the coming week, figure out the staff actions needed to support those events, and identify issues that would come up in the NAC. Tuesdays were devoted to gaining intelligence on issues from the NAC Ambassadors luncheon that were likely to come up at the formal NAC meeting on Wednesday. On Wednesday, the CMC would brief the NAC on military affairs. On Thursday, the CMC would chair the NATO Military Committee (MC). Fridays were devoted to turning out taskers to SACLANT and SHAPE on requirements or issues that had occurred at the NAC or the MC earlier in the week.

The CMC during my time in Brussels was Admiral Guido Venturoni of the Italian Navy. The Admiral had been the Chief of Defense for Italy. Lt Gen Byron developed a close private working relationship with the Admiral. Admiral Venturoni had a lot of experience in political/military relations at the highest level, and was very cordial to work with. The DCMC would often go to the CMC's office for private discussions. He would also draft private memorandums to the CMC on specific topics that outlined the current issues, provided in depth analysis as well as substantive recommendations.

The DCMC position was one of the U.S. Chairman of the Joint Chiefs of Staff's so called twelve apostles. There were jobs that the CJCS personally controlled. The DCMC was one of those jobs. The U.S. had agreed that we would never nominate an officer for the Chairman of the Military Committee if we could have the Deputy Chairman of the Military Committee on a permanent basis. The DCMC also had special responsibilities covering Nuclear, Biological and Chemical matters that required a U.S. officer with appropriate security clearances.

The roles of the DCMC and the U.S. MILREP to the Military Committee are completely separate. The DCMC is an honest broker: he does not have to provide the CMC with the U.S. position on a given issue, or automatically support U.S. recommendations. The DCMC is responsible for providing the CMC with his best military advice, which may not always agree with the U.S. position. The DCMC would always do this in private with the CMC. On the other hand, the U.S. MILREP always provides and supports the U.S. national position before the MC.

The relationship between the DCMC and the U.S. MILREP for the majority of the time while I was the MA was not good. The DCMC had held the MILREP's job prior to accepting the DCMC position. The DCMC had not said anything to me about his successor, but I soon learned that their relationship was civil at best. I learned this when I made my first office call with the Deputy U.S. MILREP. When I did my introductory office call with the General, he asked me what my job was. I gave him the billet description of what I was supposed to do. He shook his head and said, "our jobs are to keep our respective bosses out of each other's way as much as possible". So, I tried to make sure that the Deputy MILREP and I were on the same page in terms of supplying information to our respective bosses.

Under the provision of no surprises, I used to let one of the members of the MILREP's staff read the CMC's proposed speaking notes before the MC meetings. At one point, I started supplying him with copies of the notes. These were not the finalized version of what the CMC would say at the MC and so, there were often changes that would occur between our pre-briefing before the MC with the CMC, and what the CMC would actually say at the MC meeting. That exchange of information on the CMC's speaking notes stopped the day I looked over and saw the U.K. MILREP reading the U.S. MILREPs briefing book, which included the CMC's draft comments for the MC.

It was interesting to see the personal dynamics on the MC. You would think that the U.S. MILREP would always be the most dominant individual, owing to the U.S. military's overwhelming power and capabilities. However, that is not always the case. The credibility of the individual MILREPs is based on their military expertise and personal relationships with their fellow MC members, not necessarily their countries military capability. Some MILREP's punched way above their weight based on their countries relative military capability. During my time as the MA, the German and Turkish military representatives were the most dominant individuals on the MC. The German MILREP could be counted upon to make sure that whatever was being proposed to the MC was following established NATO Policy and procedures as outlined in MC Documents. As you might expect, the Germans were strict about following the rules and regulations. The Turkish MILREP was respected for his overall demeanor and experience in military operations. If the Turkish MILREP agreed to an operational plan or proposed Military Committee document, then the other MILREPs from the smaller nations believed that the plan or proposal made sense. Turkish concurrence provided top cover for the MILREP's from some of the smaller military members to go along with whatever was proposed.

My daily routine did not really change from the time I had spent as EA to the COS at AFSOUTH. I arrived early, reviewed the message traffic, proposed documents, daily schedule, and press clippings. Then I would wander down to the DIMS office and talk with his MA. My next stop was to see the MA to the CMC. When the DCMC arrived, I would give him my five-minute brief. The brief included everything that happened over the last 24 hours, daily schedule, what issues the MA's indicated the CMC and the DIMS were concerned with. Then the DCMC would provide me with directions and guidance on what he needed to prepare for the meetings and briefings he would be attending on that day.

Often LtGen Byron would have me plant ideas or suggestions as trial balloons with the other MA's, who would then discuss them with their respective principals. If I gave the other MA's the DCMC's thinking or position on a particular issue, it was because I had been instructed to do so by the DCMC ahead of time. The other MA's would also give me a hint on what their bosses were thinking on a particular subject and I would take it back and talk to the DCMC about it. This routine was to ensure that there were no surprises and that no one was embarrassed or blind-sided at a meeting. That way, whether in a meeting or briefing, all the principals had a fairly good idea on what the others position was on a given issue.

The CMC would brief the NAC in their formal meetings on Wednesday and supply a report on military activities by the

Alliance. He would give them an idea of documents, plans, or operations that the MC was working on, and what taskings had been sent to SACLANT or SHAPE. The CMC would then take questions and either provide a response or promise an update to the NAC or the individual Ambassador that had a question or concern. The DCMC would also attend the NAC and occasionally provided the brief if the CMC was not available. I was a back bencher who took notes and would review them with the DCMC when we returned to the office. LtGen Byron would also frequently do office calls with the individual members of the MC to discuss their issues or objections to documents or plans that were being considered by the MC or the NAC. He would then provide feedback to the CMC and the DIMS on how to work through the issues that the individual MILREPs would identify during those meetings. The DCMC would also occasionally talk with the Chief of Staff at SHAPE on operational planning issues and give informal feedback on what the mood in the NATO HQ was like on any given issue.

Getting consensus among all the NATO nations on any given issue was hard. But there is tremendous value and strength in that consensus once it has been achieved. There were times when a MILREP was forced to defend his nation's political position on a given issue when it did not coincide with the best military advice. This was especially difficult when a single nation was refusing to join the consensus on a given issue, thereby guaranteeing that NATO as an alliance would not act. These issues could be settled by agreeing to a "coalition of the willing." Nation X did not necessarily have to take part or provide support to a particular operation, but they did not object to the operation taking place under NATO leadership.

On September 11, 2001, I had gone for a run outside NATO HQ. When I returned, I walked down a hallway to the operations division. I almost never did this. When I walked into the OPS office

I saw that the staff were all focused on the television. Like most Americans that day, I could not believe my eyes when I saw the replays of the planes striking the twin towers. I went up to our office and talked briefly with the DCMC. At this point, VADM Malcolm Fages had replaced LtGen Byron as the DCMC. No one in OPS could tell if these were isolated incidents or part of a larger terrorist campaign. We expected the U.S. to go to a higher DEFCON level and talked a little bit about how U.S. personnel serving in Brussels would be affected. I excused myself and called my mother-in-law who was living outside of Washington, D.C. My oldest daughter Carey worked for American Airlines as ground staff but was known to take off on space available flights from time to time. A couple of the planes used on 9-11 were from American Airlines. I was concerned for her safety and she was living with her grandmother at the time. When my mother-in-law answered the phone, the first thing she said was that Carey, my daughter, was asleep in her room and had worked earlier that day. Then she told me that the Pentagon had been hit. I began to worry about our other daughter Kelly, who worked in downtown D.C. for the Fraternal Order of Police, and what might happen to her. Like thousands of other Americans in the Washington area, Kelly's trip home from work that day would be long, unusual, and memorable.

I was brought back to the present situation when a representative from the U.S. Mission called and asked what protection the DCMC would need in light of the terrorist attacks. I asked if there had there been any increase in specific threat warnings about U.S. Military personnel in Europe. I was dual hatted as the MA to the DCMC and also as Chief of Staff for U.S. Military personnel serving on the International Military Staff. I wanted to be able to give our folks guidance on what the current situation was and any added security precautions they should take. I said I would talk with the DCMC and see if/what he wanted by way of added protection.

VADM Fages wanted to know if there were any specific threat warnings about U.S. personnel and I stated that we hadn't seen any yet. We talked about the possible threat to him as a senior U.S. military official in Brussels. After some discussion, the DCMC asked me to see what force protection could be provided. I talked with the U.S. Mission and we eventually decided to ask the Belgian Police to supply protection for VADM Fages residence, and personal protection on his way to and from work. The Belgian's agreed to do so and the presence of the security detail in our office became a normal thing. The security detail enjoyed working with VADM Fages because he was a regular runner and biker. The security folks were getting a good daily workout while also supplying protection for the DCMC.

The NATO HQ became a fortress, at least as much of a fortress as there can be in a major metropolitan city. There were added road blocks, security check points, gates etc. that added a bit of time to getting to work every day. Given my early start time, I was not inconvenienced very much but the lines outside the headquarters to get into work could be pretty long on some days.

The European U.S. military community had been on higher alert for terrorism activity since the African bombings in 1997; the Armed Forces Radio Television Network constantly emphasized the need for force protection. Most U.S. personnel had grown accustomed to varying their time of departure and routes they took to work as much as possible. There was also the issue of not standing out as an American as much as possible. I always thought it was amusing when I was the CO of NSA Naples and folks (generally Army or Air Force), would say how they needed to have "cover plates" for their vehicle. A "cover plate" is the same plate that any local national driver in their country would have on their car as opposed to the AFI plates that readily showed the vehicle as belonging to a member of the U.S. military. We could get them a cover plate. But I wondered if the terrorists would figure out you were U.S. military when they saw your Ford F-150 truck that had a "*Don't mess with Texas*" on the bumper sticker. Or the fact that when you got out of the truck, you put on your cowboy hat and adjusted the plate sized belt buckle on your jeans to go with the T-shirt that had the American flag on it.

VADM Fages had come from the submarine community. Even though he was a submariner, the Admiral used to wear brown shoes occasionally because he had previously served as the Chief of Staff on the USS THEODORE ROOSEVELT carrier battlegroup. He was more circumspect and introspective than his predecessor, and was less inclined to use me as a sounding board on how to work issues within the NATO staff. I was no longer able to supply as much of a heads up to my fellow MA's on what the DCMC's position was on any given issue. When the DCMC would raise an issue during a meeting with the CMC and the DIMS, the other MA's would glare at me with the implication, *why didn't you tell us he was for/against this?* The Admiral was inclined to play his cards close to the vest, and get directly to the point, as opposed to working around to it gradually as was usually the case in NATO.

I told folks that had to brief him to front load their presentation with the issue to be decided, rather than building towards it gradually. I said, "you won't get beyond two or three slides in the briefing and then the Admiral will start to ask questions. Answer his questions and forget about completing the briefing, you won't get there." Inevitably, after the officer did not get through their entire briefing, I would get a phone call from them asking what was wrong with their briefing, or what they could change? I said you have to take into account how the Admiral processes information. The DCMC would read the briefing ahead of time. He knew the material and had figured out what issues he needed more information on. He was not interested in letting you get through the entire briefing before he would start getting answers to his questions.

As the assignment in Brussels was drawing towards a close, I began to look for my final or "twilight" tour in the Navy. The Detailer opened our negotiation with the following statement. You know, every O-6 has to do a tour overseas sometime in his career and you should look at doing another overseas tour. Would you consider going to Japan for a chief of staff job? I was a little incredulous and said, so five years in Naples, and three years in Brussels doesn't count? Find me another Navy O-6 who has been overseas for eight consecutive years as a Captain. That ended the conversation about going overseas. At one point the possibility of becoming CO of the Pacific Missile Range Facility in Kauai came up, but I wasn't prepared to spend another tour in Hawaii. We needed to get back to the states to be closer to family. VADM Fages had just returned from a visit to the U.S. and said he had found the perfect "twilight tour" position for me. While in D.C., the DCMC had used the office of the U.S. MILREP and found out that the O-6 who ran the office would be rotating and they were looking for a replacement. He encouraged me to check it out and see if that might work as my final tour in the Navy.

The U.S. MILREP had an office in the Pentagon that was staffed by an O-6, E-7, and a civilian secretary. The personnel and space were a legacy of when the MILREP had been a four-star position. I wandered down from the DCMC's office to the MILREP's office in Brussels. I talked with his Chief of Staff about the position, what it entailed, where they were in their search for a successor, and whether he thought there was a chance that the MILREP would consider me for the position? Both the DCMC and the MILREP had changed during my last year as MA and the relationship between the principles and their staffs was much better. I indicated that I would be interested in the position and asked the

MILREP's COS to informally see if Lt Gen Timothy Kinnan, the MILREP, wanted to consider me for the job. The word came back that Lt Gen Kinnan wanted to see me.

Arriving at the General's office, we talked about the position, my background and experience, and why I wanted the job. It seemed like the logical conclusion to my career, especially as concerned with NATO. I would be comfortable representing his office in D.C. as the Liaison Officer to the Joint Staff. I was up to speed on all the issues affecting NATO and could be effective from the first day on the job. I had a house in a Washington Suburb and would be moving back in. Earlier Liaison Officers tended to come from the Joint Staff where they had been working on NATO issues, but none had the experience that I had of actually serving in NATO billets, especially overseas. I would have worked at every military level of the NATO Alliance. Starting with a Major NATO Commander (SACLANT in Norfolk, VA), a Major Subordinate Commander (AFSOUTH in Naples, Italy), the International Military Staff (Brussels, Belgium), and finishing with the U.S. Military Delegation to NATO as the Liaison Officer to the Joint Staff (Washington, D.C.). The MILREP said he wanted me to take the job and I went home and told the wife that we would be returning to D.C.

CHAPTER XVIII

TWILIGHT TOUR

U.S. MILITARY REPRESENTATIVE TO THE NATO MILITARY COMMITTEE

LIAISON OFFICER TO THE JOINT STAFF

WASHINGTON, D.C.

JUNE 2002 – JUNE 2004

A twilight tour is your last tour on active duty. We had mixed emotions about moving back to the U.S. As empty nesters for our last two years in Brussels, we took full advantage of the opportunity to travel. We would be returning to our family and adjusting to all the changes that had occurred during those eight years in Europe. Having spent the last seven years in close proximity to the decision makers in Naples and Brussels as an EA, CO, and MA, I would go back to being a staff officer, i.e., one at a much higher pay grade than my earlier experience on the Navy Staff.

My new routine was to get to work a little after six in the morning and contact the MILREPs Chief of Staff (COS) in Brussels. I was now on the wrong end of the six-hour time difference between Brussels and D.C. After checking in with the COS, the first event on my schedule was a morning meeting with the head of the Europe, NATO, and Russia Branch in the J-5, Joint Staff for Strategy, Plans and Policy. The one star would meet with the section chiefs from Europe, Russia, and NATO. I was there to provide a European context and protect the interests as well as the position of the MILREP in Brussels.

The job was a holdover from the time before e-mail, fax, and improved communications allowed the action officers on the J-5 Staff to talk directly with their counterparts in the U.S. Delegation in Brussels on a real time basis. Supplying context for the discussions that took place in D.C. to the COS and MILREP in Brussels was my job. As for the folks on the J-5 staff, I knew all the Brussels players personally and could give them suggestions on how other members of the NATO Military Committee would react to potential U.S. proposals. I still thought of myself as more of a NATO staff officer than a U.S. Navy staff officer; as a result, I would not hesitate to tell them when I thought the U.S. position was not the best possible course of action. I would always preface those statements with, either "this is my personal opinion," or based on what they are hearing in Brussels "the MILREP/COS recommends that the U.S. position should be the following."

Following the morning meeting I would wander around the J-5 spaces and see what was going on. I would also talk to the folks who worked in the Office of the Secretary of Defense's NATO shop. The Joint Staff and OSD were not always on the same page and it was interesting to see the different perspectives and how they were represented. I would then call Brussels and give the COS a back brief on the morning's activities. He would supply me direction on issues that needed to be clarified and who he wanted me to contact. Then I would go to the gym. The good thing about starting at 0600 in the morning was that I had done all of the above by one o'clock. That left me time to go to the gym, clean up any issues, and provide an end of day summary to the COS that he would read when he came to work the next morning in Brussels.

We were the one stop shop for NATO in D.C. OSD and the Joint Staff would steer folks our way that were looking for more information on NATO. I also gave lectures to U.S. folks who were about to be assigned to a job in NATO at the NATO Staff Officers Course which was taught at the National Defense University. I talked about the Battle Rhythm of the Headquarters in Brussels and the interrelationship of the different military echelons in NATO. Having served at every military level of the alliance, I was able to provide them with a perspective on the different NATO commands, their issues, and how to be as effective as possible in their tour as a NATO Staff Officer. I told them that the most important thing they had was their credibility and military professional expertise. They needed to develop a reputation for telling the truth and not merely parroting the U.S. national position. Their value and effectiveness as a staff officer was related to their supplying the best military advice, and not the best political advice. If their staff contemporaries from other NATO countries sought their advice and counsel on tough issues, it was a sign that they had credibility and in the long run could more effectively represent the United States.

The folks I taught were shocked that I had never attended their course prior to my assignment to SACLANT, or any of my later NATO tours. There wasn't time when I was going initially to SACLANT to attend the course, and after that tour in Norfolk, I felt that I would not learn anything new by attending the course prior to later NATO assignments. The only NATO course I ever attended was the Crisis Management Staff Officers Course in Oberammergau, Germany. Later, I would be invited back to that course to lecture on the changes in the NATO Precautionary System, NATO Rules of Engagement, and Military Response Options in Crisis, and how they related to the development and implementation of operational planning.

In retrospect, I probably should have tried to get a NATO job in OSD, Joint Staff or on the Navy Staff for my twilight tour. My qualifications were excellent and working on those staffs would have better prepared me for the coming transition to civilian life. I would have been in a better position to begin networking, which would have provided me with a better opportunity to transition into civilian life. The reason I did not try for any of those positions was that I was burned out professionally. I was not up for the daily grind of negotiating policy between OSD, the Joint Staff, and the Military Delegation in Brussels. I was happy being an over qualified action officer who could leave the job at the end of the day and not feel like there was something that I had missed.

That feeling persisted through the first year of my time as the Liaison Officer, but I began to feel like I had sold myself short during the second year. It became difficult to watch decisions being made by folks who did not have my experience and background. No longer a key player in the room, I was just a bystander who could not decisively influence the decisions being made and began to regret having settled for a less influential position. People would humor me and listen to what I had to say, but I wasn't making a difference or having the impact that I had been accustomed to in my earlier positions.

While I always knew there were places and positions within the Navy that showed how well you were doing in your career, I never had a plan that at a certain point I would have a specific job on a specific staff in a specific location. When it was time to rotate, I would consider what jobs I was being offered and choose the one that was best for me personally and professionally. For me, I was always afraid that when I started thinking about my next job, I would not give 100% to the job I was currently holding. I would spend too much time thinking about what it would be like in the new job and lose the focus and effectiveness in doing my current job to the best of my ability. As a result, I did not think a lot about what I would do when I retired and did not set myself up for a smooth transition. I was not prepared to begin actively looking for my next job while still on my current job and did not spend any time networking to set up my transition.

I did not have a retirement ceremony although I had often encouraged my contemporaries to do so as a way to reflect on their service while bringing an end to that part of their lives. It did not make a lot of sense to have a ceremony in either D.C. or in Brussels -I did not want to inconvenience the folks in either place. The people in Brussels had not worked with me on a daily basis over my final two years. The folks in D.C. were not a part of my organization and it would have been unfair to lean on them to come up with a ceremony. They both asked me about whether I wanted a ceremony and were willing to do whatever was necessary to send me over the side with proper honors, but it did not feel right. There was no big ceremony welcoming me into the Navy when I enlisted in 1973, and none was necessary when I retired in 2004. I went to Brussels for a courtesy call with the U.S. MILREP and then through the check-out process with the Joint Staff and the U.S. Navy.

On 1 July 2004 I was a civilian again.

POST SCRIPT

I wrote this book for my wife and daughters, to give them an idea of what was happening when I left for work and during those periods when I was deployed away from home. I wanted to give them an idea of what I was doing and thinking during particular periods in their lives.

This memoir is also a way of documenting the journey of a boy from Cleveland's West Side who becomes a Naval Aviator, sees the world, and achieves the significant career milestones of becoming the commanding officer of a Navy Helicopter squadron, Base Commander of one of the most important Navy installations in Europe, and five years as an EA/MA at the highest levels of the NATO military alliance.

The U.S. Navy trained me to be a Naval Aviator, taught me to think outside the narrow confines of my education and upbringing, and allowed me to explore the world. Service in the Navy expanded my horizons both personally and professionally. In return, I gave the U.S. Navy thirty years of my life and got the better end of the deal.

I was a part of the expanding missions and capabilities of the Navy Light Airborne Multi-Purpose System (LAMPS) helicopter community and one of the initial Joint Staff Officers, who spent more time in Joint Duty Critical Billets than any of my contemporaries in the Navy, including eight consecutive years overseas. I was a participant in the cold war, the expansion of NATO, a Commanding Officer during periods of combat operations in the Persian Gulf and Kosovo, and a witness to the start of the war on terrorism.

The thing about being a Commanding Officer is that you never really know what effect you are having on your people. You want to believe that you did the right things and that the command was a better place for your having been there. But is that reality or just a dream?

It was not until I attended an H-2 reunion in October of 2018 that I got an appreciation of what I had accomplished. I was standing in the bar hydrating because we were in Pensacola and you can get dehydrated there very easily, even in October. A series of faces that I had not seen in 27 years approached and every one of them started the conversation with some variation of, "I bet you don't remember the time when you were the Skipper and you told me (fill in the blank)". It could have been about leadership, taking care of your people, knowing the aircraft and your emergency procedures cold, or even the "you have to go out, but you don't have to come back" speech. Inevitably they had carried those memories around with them for the rest of their lives, and they were positive experiences.

The most complimentary comment came from an aviator I had gone through the FRS with when he was an Ensign. He later commanded a squadron of his own. Talking about his days in HSL-37 when I was the Maintenance Officer and he was on detachment he said, "all the JO's in the squadron knew that you were the real deal and respected you."

Respect, it does not get any better than that!

History of Assignments

Naval Aviation Training Command, September 1973-December 1974

Pensacola, FL, Aviation Officer Candidate School, Student Naval Aviator, Saufley Field (VT-1), Whiting Field (VT-6, HT-8, HT-18)

Naval Air Station Bermuda, January 1975 – October 1977

Saint Georges, Bermuda, Search and Rescue Pilot, Avionics Division Officer, Assistant Communications Officer, Manager Commissioned Officers Mess Open, Fire House Division Officer

Helicopter Anti-Submarine Squadron Light Three Zero Detachment Alpha, March 1978 – January 1980, Detachment Administrative, Operations, Maintenance, Division Officer, USNS HARKNESS (TAGS-32), Eastern Mediterranean, Northern and Southern Approaches to the Suez Canal, 1st Lieutenant, Public Affairs Officer

Helicopter Anti-Submarine Squadron Light Three Four, January 1980 – June 1982

Norfolk, VA, Assistant Training Officer, Maintenance Officer Detachment One USS PETERSON DD-969 Persian Gulf September 1980 – March 1981, Officer in Charge (OINC) Detachment Four USS CONNOLE FF-1056 Mediterranean Sea June 1981- November 1981, Naval Air Training and Operating Procedures Standardization (NATOPS) Officer, Assistant Safety Officer

Office of the Chief of Naval Operations (OPNAV), June 1982-February 1985

Washington, D.C., General Planning and Programming Division, OP-90, OP-906E6 Assistant Plans and Policy Staff Officer, Congressional Special Interest Correspondence Section

Helicopter Anti-Submarine Squadron Light Three Seven, August 1985 – February 1988

Barbers Point, Hawaii, OINC Detachment Four USS BAGLEY FF-1069 Western Pacific, Indian Ocean, Mediterranean, January 1986 – June 1986, Training Officer, Maintenance Officer

College of Naval Command & Staff, March 1988 – March 1989

Newport, Rhode Island, Student

Helicopter Anti-Submarine Squadron Light Three Seven, May 1989 – June 1991

Barbers Point, Hawaii, Executive/Commanding Officer

NATO Supreme Allied Commander Atlantic, July 1991 – August 1994

Norfolk, VA, Operations, Crisis Management Staff Officer

NATO Allied Forces Southern Europe, August 1994 – March 1997

Naples, Italy, Policy Branch Head, Executive Assistant to the Chief of Staff AFSOUTH, Executive Assistant to the Chief of Staff IFOR, NATO Implementation Forces Sarajevo, Bosnia Hercegovina

Naval Support Activity Naples, Italy, March 1997 – June 1999

Naples, Italy, Commanding Officer

NATO International Military Staff, June 1999 – June 2002

Brussels, Belgium, Military Assistant/ Chief of Staff to the Deputy Chairman NATO Military Committee (DCMC)

U.S. NATO Military Committee Representative to the Joint Staff, June 2002 – 2004

Washington, D.C., Liaison Officer

FLIGHT TIME

AIRCRAFT	HOURS	TIME FRAME
T-34 Saufley Field	29.8	6 Dec 73 – 5 Feb 74 VT-1
T-28B/C Whiting Field (North)	96.5	14 Feb 74 – 25 June 74 VT-6
TH-57A Whiting Field (South)	28.5	1 July 74 – 1 August 74 HT-8
TH-1H/L HT-18 Whiting Field (South)	66.3	7 August – 11 December 74
UH-1N NAS Bermuda	500.6	13 January 75 – 5 October 77
HH-2D/SH-2F HSL-30, HSL-34, HSL-37	2,645.4	2 November 77 – 12 June 91
SH-3A Flight Co-Pilot (HSL-30)	3.5	12 February 79 VRF-30 Ferry
SH-60B Barbers Point, HI (HSL-37)	1.0	2 February 87 HSL-45 @
C-12 NSA Naples, Italy	310.3	20 November 97 – 1 June 99
Fixed Wing	447.6	
Rotary Wing	3234.3	
Total Flight Time	3681.9	

Decorations, Medals, Badges, Citations, Campaign Ribbons

Defense Superior Service Medal (3)

Legion of Merit

Defense Meritorious Service Medal

Meritorious Service Medal

Joint Service Commendation Medal

Navy & Marine Corps Commendation Medal (3)

Joint Service Achievement Medal

Joint Meritorious Unit Award (2)

Meritorious Unit Commendation (2)

Navy "E" Ribbon (2)

Navy Expeditionary Medal

National Defense Service Medal (3)

Armed Forces Expeditionary Medal

Kosovo Campaign Medal

Global War on Terrorism Service

Armed Forces Service Medal

Humanitarian Service Medal

Navy Sea Service Deployment Ribbon (3)

Navy & Marine Corps Overseas Ribbon (9)

NATO Medal

Aviator Insignia

Command Ashore Pin

Command At Sea Pin

ABOUT THE AUTHOR

Jack Coyne, Naval Aviator, Author

Jack is a retired Naval Aviator with over thirty years' service whose career included flying in the Atlantic, Mediterranean, Middle East, and Western Pacific. He also served in senior staff positions at every military level of the North Atlantic Treaty Organization (NATO), including Norfolk, Virginia, Naples, Italy, Brussels, Belgium, and Washington, D.C.

He commanded Helicopter Anti-Submarine Squadron Light THREE SEVEN at Barbers Point, Hawaii during OPERATION DESERT STORM in Kuwait, and Naval Support Activity Naples, Italy during OPERATION ALLIED FORCE in Kosovo.

An unrestricted Naval Aviator, he initially flew UH-1N helicopters doing Search and Rescue missions in Bermuda. He then transitioned to the SH-2F in the Light Airborne Multi-Purpose System (LAMPS) community doing anti-submarine warfare, anti-ship surveillance and targeting, and utility missions including Search and Rescue, Vertical Replenishment, and medical evacuations while assigned to squadrons in Norfolk, Virginia, and Barbers Point, Hawaii. He finished his aviation career flying fixed wing C-12's as a utility pilot at NSA Naples, Italy.

One of the initial Joint Staff Officers, he served in a number of joint duty billets in NATO including: Military Assistant to the Deputy Chairman of the North Atlantic Treaty Organization (NATO) Military Committee, in Brussels, Belgium; Executive Assistant to the Chief of Staff at Allied Forces Southern Europe in Naples, Italy, as well as for the Implementation Force (IFOR) in

Sarajevo, Bosnia Hercegovina; Policy Branch Chief, Allied Forces Southern Europe in Naples, Italy; the U.S. NATO Military Representative's Liaison Officer to the Joint Staff in Washington, D.C.; and as the Crisis Management Staff Officer at Supreme Allied Commander Atlantic, Norfolk, Virginia. He also served on the Chief of Naval Operations Staff in OPNAV-906, Congressional/Special Interest Correspondence section in Washington, D.C.

His early career was focused on broadcast journalism. Prior to joining the Navy he covered news and sports while working at radio stations WOUB, WLGN, and WATH in Southeastern Ohio, and for Public Broadcasting System television stations WOUB in Athens, Ohio, and WVIZ in Cleveland, Ohio. While in the Navy he authored a weekly newspaper column and hosted a weekly radio program while serving as the Commanding Officer of Naval Support Activity Naples, Italy.

His academic credentials include a Bachelor of Science in Communications (Radio/Television) from Ohio University, a Master of Arts in Government (National Security Studies) from Georgetown University, and a Master of Arts in National Security and Strategic Studies from the Naval War College in Newport, Rhode Island.

"A Cold War Sailor" represents his first published book.

Made in United States
North Haven, CT
10 November 2026

22091337R00183